AF584856

Pearson Australia
(a division of Pearson Australia Group Pty Ltd)
459–471 Church Street, Level 1, Building B, Richmond, Victoria 3121
PO Box 23360, Melbourne, Victoria 8012
www.pearson.com.au

First published 2016 by Pearson Australia
2026 2025 2024 2023
10 9 8 7 6 5 4 3 2 1

Publisher: Alicia Brown
Development Editor: Shirley Melissas
Project Manager: Shelly Wang
Production Manager: Elizabeth Gosman
Editors: Deborah Lombardo, Aptara
Proof Reader: Vicky Chadfield
Designer: Anne Donald
Copyright & Pictures Editor: Samantha Russell-Tulip
Desktop Operators: David Doyle, Aptara
Illustrator: DiacriTech
Printed in Singapore by Markono/07 (2024)

ISBN 9781488615061

Pearson Australia Group Pty Ltd ABN 40 004 245 943

Disclaimer/s
The selection of internet addresses (URLs) provided for this book/resource was valid at the time of publication and was chosen as being appropriate for use as a secondary education research tool. However, due to the dynamic nature of the internet, some addresses may have changed, may have ceased to exist since publication, or may inadvertently link to sites with content that could be considered offensive or inappropriate. While the authors and publisher regret any inconvenience this may cause readers, no responsibility for any such changes or unforeseeable errors can be accepted by either the authors or the publisher.

Some of the images used in *Pearson Science* might have associations with deceased Indigenous Australians. Please be aware that these images might cause sadness or distress in Aboriginal or Torres Strait Islander communities.

Practical activities
All practical activities, including the illustrations, are provided as a guide only and the accuracy of such information cannot be guaranteed. Teachers must assess the appropriateness of an activity and take into account the experience of their students and facilities available. Additionally, all practical activities should be trialled before they are attempted with students and a risk assessment must be completed. All care should be taken and appropriate protective clothing and equipment should be worn when carrying out any practical activity. Although all practical activities have been written with safety in mind, Pearson Australia and the authors do not accept any responsibility for the information contained in or relating to the practical activities, and are not liable for any loss and/or injury arising from or sustained as a result of conducting any of the practical activities described in this book.

Acknowledgements
We would like to thank the following for permission to reproduce copyright material. The following abbreviations are used in this list: t = top, b = bottom, l= left, r = right, c = centre.

Cover image: Science Photo Library/Eric Heller

Alamy Stock Photo: Graphic Science, p. 56.

Avita Medical Ltd.: p. 49r.

Department of the Environment and Energy: Commonwealth of Australia, used with permission from energyrating.gov.au, p. 74.

Fotolia: Petrol, p. 76 (zebra); Ilona Pitkin, p. 76 (disco ball); Slava, p. 51; Taras Vyshnya, p. 124.

Getty Images: Tony Ashby, p. 47; Edward Kinsman, p. 25(ballpoint pen); MedicalRF.com, p. 57r; Oxford Scientific, p. 55; Photo Researchers, p. 61l; Science Photo Library, p. 49c.

iStockphoto: Eraxion, p. viii(human body).

Pearson Education Ltd.: p. 35.

Pearson Education, Inc.: pp. 75(all), 76(mirror, magnifying glass, hand mirror, camera, binoculars, eye, glasses).

Science Photo Library: p. 64; Dr Jeremy Burgess, p. 25 (salt crystals); Thomas Deerinick, p. 25 (fly eye); Eye of Science, p. 25 (cotton fibres, rusty nail); Eric Heller, p. i; Medi-Mation, p. viii (heart); Sidney Moulds, p. ix; Susumu Nashinaga, p. 25 (sunflower pollen); Dr Yorgos Nikas, p. 61c; Gary Parker, p. 49l; David Scharf, p. 25 (paper fibres); Power and Syred, p. 25 (fly foot).

Shutterstock: Sergey Furtaev, p. 76 (microscope).

Thinkstock: Elnur Amikishiyev, p. 72 (vacuum); Rudolf Balasko, p. 72 (globe); Kim Reinick, p. 72 (toaster); tiridifilm, p. 72 (TV).

University of New South Wales: Based on: Harris K, Fitzgerald O, Paul RC, Macaldowie A, Lee E & Chambers GM 2016. *Assisted reproductive technology in Australia and New Zealand 2014*. Sydney: National Perinatal Epidemiology and Statistics Unit, the University of New South Wales, p. 65.

Wake Forest Institute for Regenerative Medicine: p. 32.

Every effort has been made to trace and acknowledge copyright. However, should any infringement have occurred, the publishers tender their apologies and invite copyright owners to contact them.

Contents

How to use this book • ACTIVITY BOOK

Pearson Science 2nd edition Activity Book

An intuitive, self-paced approach to science education, which ensures every student has opportunities to practise, apply and extend their learning through a range of supportive and challenging activities.

Pearson Science 2nd edition has been updated to fully address all strands of the new Australian Curriculum: Science, which has been adopted throughout the nation. This edition also captures the coverage of Science curricula in states such as Victoria, which have tailored the Australian Curriculum slightly for their students.

The *Pearson Science 2nd edition* features a more explicit coverage of the curriculum. The activities enable flexibility in the approach to teaching and learning. There are opportunities for extension as well as reinforcement of key concepts and knowledge. Students are also guided in self-reflection at the end of each topic.

Explicit scaffolding makes learning objectives clear and includes regular opportunities for reflection and self-evaluation.

In this edition, we provide a structured approach that integrates a seamless, intuitive and research-based learning hence **differentiating** the course for every student.

The Activity Book also provides richer application opportunities to take the Student Book content further with explicit coverage of Inquiry Skills, Science as a Human Endeavour and Science Understanding.

The diverse offering of worksheets allows students to be challenged at their level. Students have the flexibility to be self-paced and this new edition comes with the advantage of each worksheet being self-contained.

Be guided

A new handy **Toolkit** at the beginning of the Activity Book has been created to build skills in the key areas of practical investigations, research, thinking, organising, collecting and presenting. Each skill developed in the toolkit is directly relevant to applications in questions, investigations and research activities throughout the student and activity books. A toolkit spread provides guides and checklists alongside models and exemplars.

Be supported

Vocabulary boxes provide definitions for key terms, within the relevant context of the task. **Hints** help students get started on a worksheet and provide support in overcoming a barrier.

Be reflective

The **Thinking about my learning** feature provides the opportunity for self-reflection and self-assessment. It encourages students to look ahead to how they can continue to improve and assists in highlighting focus areas for skill and knowledge development.

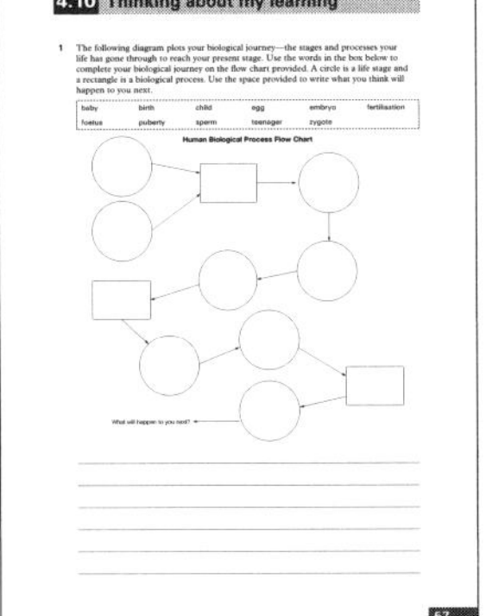

Be ready

A **knowledge preview** at the beginning of every chapter, activates prior knowledge relevant to the topic, providing an opportunity for students to show what they currently know. This handy tool supports teachers in assessing students' prior knowledge.

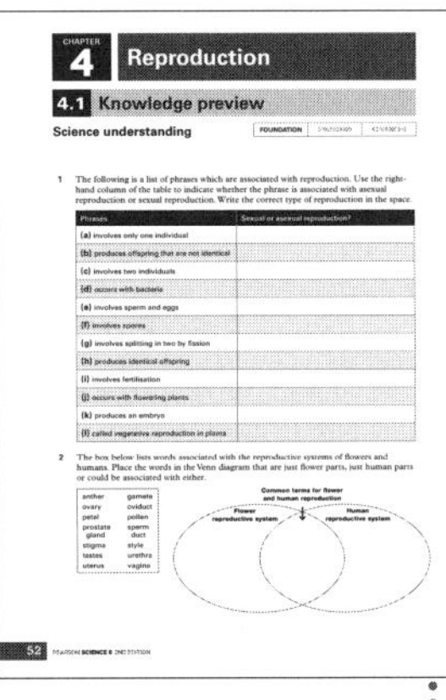

Be literate

Newly improved **literacy reviews**, in consultation with our Literacy Consultant Dr Trish Weekes, provide a deeper and broader range of language building tasks. Every chapter concludes with a literacy review which focuses on building a deeper understanding of key terms supporting students to correctly apply key terms from the topic.

Be set

Visit www.pearsonplaces.com.au to enjoy the benefits of the following digital assets and interactive resources to support your learning and teaching:

- Interactive activities and lessons
- Untamed Science videos
- Student investigation templates and teacher support
- Answers for questions in Student and Activity Book
- Chapter tests with answers
- SPARKlabs
- Weblinks
- Risk assessments
- Teaching programs and curriculum mapping audits

Worksheets

Each worksheet is classified according to the degree with which it deals with curriculum understandings.

- **Foundation** indicates the focus is on the basics like terminology.
- **Standard** indicates a focus on the core ideas, understandings and skills.
- **Advanced** indicates transfer and extension of core science understanding and skills to new or more sophisticated situations.

Teachers may use this tool to **differentiate** the worksheets allocated to students. They may select worksheets for students based on whether basic, core or extension exercises are required. The categories do not indicate the degree of difficulty of tasks on the worksheet. A worksheet labelled advanced may have tasks ranging from lower level through to higher-level thinking.

4.7 Reproduction, gestation and size

Science inquiry skills

FOUNDATION | STANDARD | **ADVANCED**

Processing & Analysing | Evaluating

The main science **strand** is identified for each worksheet: Science Inquiry Skills, Science Understanding, Science as a Human Endeavour.

Science Inquiry Skills relevant to the worksheet are identified: Questioning and Predicting, Planning and Conducting, Processing and Analysing, Evaluating, Communicating.

Each question in a worksheet is identified according to the degree of difficulty. Bloom's taxonomy is used as the basis for question classification. The classification is intentionally subtle and unobtrusive. This tool may be used to differentiate between students, matching questions to their levels of ability.

Questions with a straight number indicate a remembering or understanding lower-order question.

Questions with a circle around the question number indicate an analysing or applying middle-order question.

Questions with a square around the question number indicate a creating or evaluating higher-order question.

1 Read through the words in the vocabulary

(a) List the words that you know and wr

(5) Identify the solutes that did not dissolve in

[7] Discuss how this story illustrates that scie

An innovative tool for students to quickly and easily reflect on their understanding of each worksheet. The teacher may use the student responses as a formative assessment tool. At a glance, teachers may assess which topics and which students need intervention for improvement.

RATE MY UNDERSTANDING
Shade the face that shows your rating

PEARSON Australian Curriculum Writing and Development Team

Anna Bennett
Differentiation consultant, Victoria

Ian Bentley
Former Head of Science, VCE exam and trial exam writer, STEM investigation developer, Victoria

Christina Bliss
Science and senior Biology teacher, VCE assessor, Biology author and question writer, Victoria

Donna Chapman
Science laboratory technician, Safety consultant, Victoria

Dr Warrick Clarke
Curriculum writer, Science communicator and Australian Post-Doctoral Research Fellow at UNSW, Author, New South Wales

Jacinta Devlin
Science and senior Physics teacher, author, Victoria

Julia Ferguson
Education Officer Earth Science Western Australia, author and reviewer, Western Australia

Bob Hoogendoorn
VCAA exam assessor in Chemistry, Former senior Chemistry teacher exam-style question author, Victoria

Penny Lee
Science laboratory technician, safety consultant, Victoria

Louise Lennard
Head of Science, former industrial scientist, author and STEM investigation developer, Victoria

Greg Linstead
Former Head of Science, Education Department of WA, curriculum writer, author, Deputy Chief Examiner, STAWA President, Western Australia

Bryony Lowe
Director of Numeracy Improvement. Former Head of Science, Region Teaching & Learning Coach, author and reviewer, Victoria

David Madden
Science Learning Area Manager at QCAA, former Head of Science, author and reviewer, Queensland

Fran Maher
Bioscience educator, Science teacher, former Bioscience researcher author, Victoria

Rochelle Manners
Science and Mathematics teacher, Teacher Companion coordinating author, Queensland

Shirley Melissas
Teacher librarian and author, Development editor, Victoria

Tamsin Moore
Science and senior Psychology teacher, author, Western Australia

Natalie Nejad
Head of Science, Science and Mathematics teacher, STEM investigation developer, Victoria

Malcolm Parsons
Education consultant, former teacher, author, Victoria

Greg Rickard
Teacher, former Head of Science, author, Victoria

Lana Salfinger
Teacher, Head of Science, IB Workshop leader in MYP Sciences author, Western Australia

Maggie Spenceley
Former teacher, curriculum writer Queensland Studies Authority, author, Queensland

Jim Sturgiss
Teacher, former coordinating analyst (NSW Department of Education) reporting NAPLAN, senior test designer for Essential Secondary Science Assessment (ESSA), Thinking Scientifically questions author, New South Wales

Craig Tilley
STAV trial exam coordinator, VCAA exam assessor, former senior Physics teacher, author and exam-style question writer, Victoria

Jo Watkins
Chief Executive Officer at Earth Science Western Australia, author and reviewer, Western Australia

Dr Trish Weekes
Science literacy consultant, New South Wales

Rebecca Wood
Science educator and tutor, author, Victoria

Thinking scientifically

Science visuals

Visuals are very important in communicating science information. Visuals come in many forms: two-dimensional visuals like diagrams, photographs and video and three-dimensional visuals like models.

Cycle diagrams

The cyclical diagram is a useful two-dimensional tool for showing a process or life cycle. When explaining the meaning of a cycle diagram, you need to be able to identify the main events or processes in the cycle. Cycle diagrams demonstrate how things relate to each other and the repetitive nature of the process as in Figure 0.1.

Features of cyclical diagrams		
Title	**Labels**	**Arrows**
The title indicates what information is shown.	Labels, notes or a key can describe and identify features. The information is concise and quick and easy to understand. Note that pointers, not arrows, are used to label parts of the diagram.	Arrows are used to show processes, not to label parts and features. The widths and colours may be important in understanding diagrams.

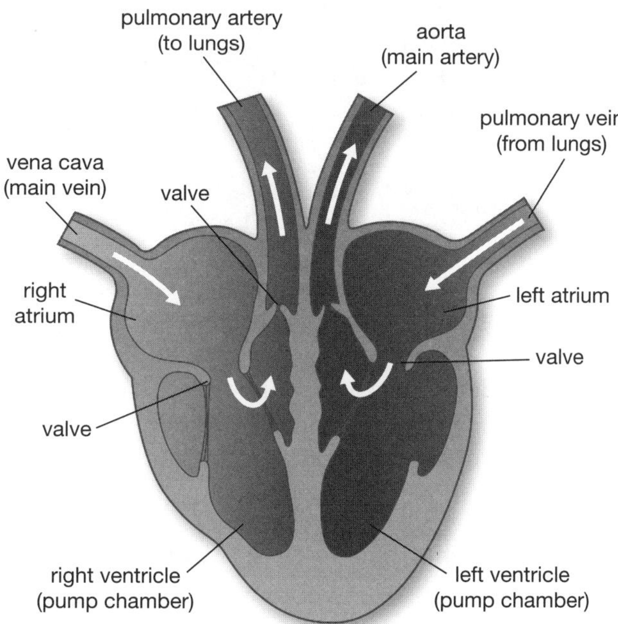

Figure 0.1 A cyclical diagram showing blood circulation in the heart

Models

Models are visuals used to help us understand complex ideas, explanations, patterns or processes. There are many different types of models. The more common models that you would see are visual and physical models; however, there are also mathematical and computational models. Models can be made (like a plasticine model of a heart), drawn or even created on a computer.

Physical models are three-dimensional and can either be larger than the object they are representing (scaled up) or smaller (scaled down). You can hold physical models and observe them closely. By turning them around and upside down the viewer can see the object from different perspectives. Figures 0.2 and 0.3 are two examples of models of the human heart.

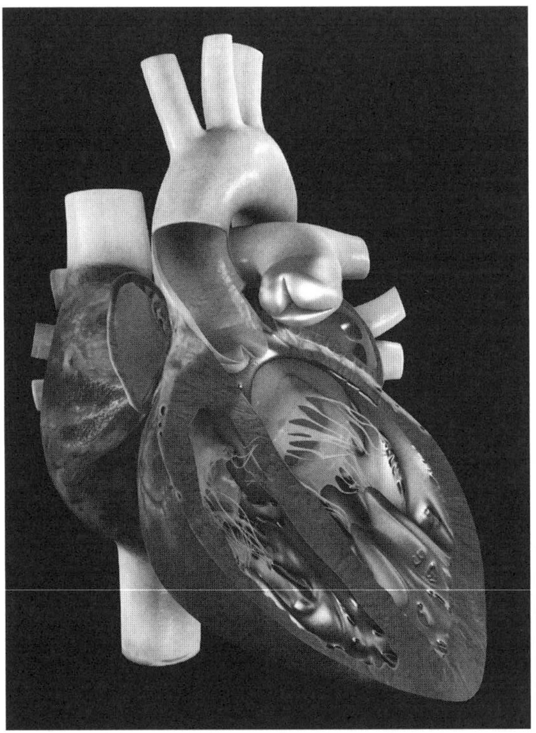

Figure 0.2 A physical model of a human heart

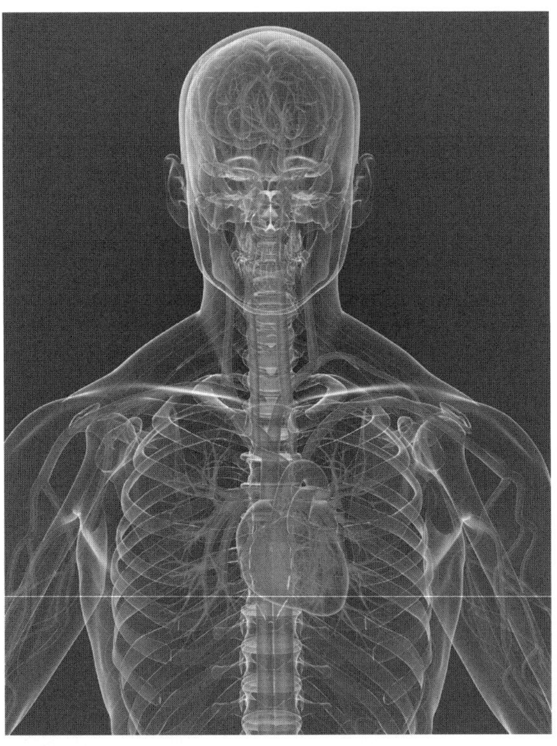

Figure 0.3 A transparent digital model of the human body with heart and blood vessels

Drawings of specimens

Drawings of specimens viewed under a microscope need to be accurate and carefully drawn. One of the most commonly viewed specimens is a cell. The following guidelines will help you draw an accurate view of a cell. Figure 0.4 provides an example of how to complete a specimen drawing of onion cells.

1. Use white, unlined paper and a sharp lead pencil for drawing. Use a compass to draw a neat circle.
2. Start by drawing a large circle on your page (at least half the page).
3. Label the circle 'field of view'. This is the entire area you can see through the microscope.
4. In the field of view circle on your page, draw the outline of the main features you are viewing. Use a fine line. In your drawing, try to accurately represent the size of the cell in the field of view.
5. Keep looking back at your specimen while you are drawing. When drawing from a microscope it is useful to look down the eye piece with one eye and at the drawing paper with the other—this is possible but will take some practice.

6 Do not shade the drawing but use a series of dots (stipple) or crisscross lines (cross-hatching) to show shading and features.

7 Include clear and proper labels for the features of the cell. Labels should be written outside of the field of view circle.

8 Label the drawing with the specimen name and magnification.

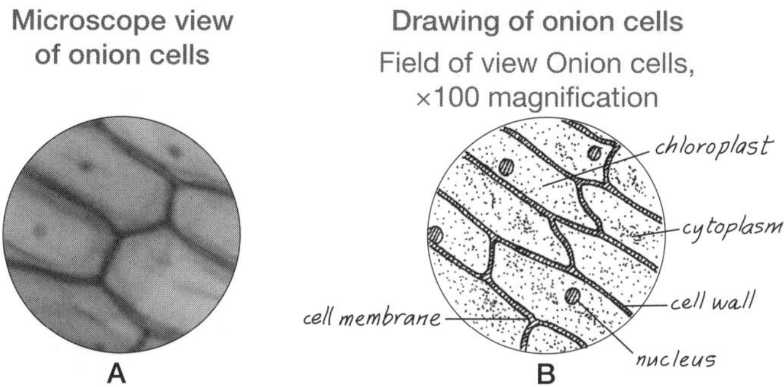

Figure 0.4 Onion cell specimen

Graphic organisers

Graphic organisers are visual displays that are used to help organise and present information or ideas. You may use graphic organisers to display ideas, for taking notes and for making summaries of information. Graphic organisers can be very helpful for learners with a preference for visual learning.

PMI charts

A PMI chart allows you to explore various aspects of a topic or idea. Using a PMI chart, you are able to examine the strengths, advantages or positives of the topic (P); the weaknesses, disadvantages or minuses/negatives of the topic (M) and the interesting features of the topic (I). Table 0.1 is an example of a PMI chart on cell cloning. Cell cloning is the process of making exact copies of a group of cells (or clones) from the original cells.

PLUS (+)	MINUS (–)	INTERESTING
• It can be used to create changes in genetic makeup of organisms to introduce positive traits and eliminate negative traits. • Organs could be produced for human transplants. Sick patients would not need to wait for an organ donor. • Infertile couples could have children without needing IVF, sperm donors or surrogacy.	• It is interfering with nature and the natural process of things. • It could lead to reduced biodiversity. • Cells that are cloned could have mutations or abnormalities, meaning all the cloned cells would too. • The cloning process still needs improvement as some organisms develop tumours or disease after the process.	• Scientists are able to clone human embryos in order to treat certain diseases. This brings with it a lot of possibilities but problems too. • Could we make exact copies of people and what would be the consequences? • There is the possibility of cloning certain animals to prevent them from becoming extinct.

Table 0.1 PMI chart on cell cloning

Venn diagrams

A Venn diagram makes it easier to compare and contrast the features or characteristics of two items or topics. A Venn diagram uses two circles that overlap. In the centre where the circles overlap, write all the common features/characteristics of the two items or topics. In the outer circles, list all the features/characteristics that are different. An example of a Venn diagram comparing plant and animal cells is shown in Figure 0.5.

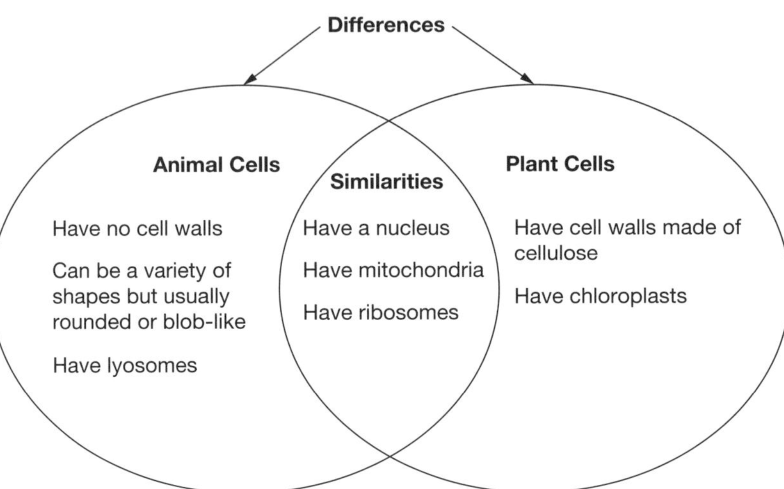

Figure 0.5 Venn diagram comparing plant and animal cells

Concept maps

A concept map organises ideas in a hierarchical branching structure. It uses words and captions. Ideas are linked by arrows. Ideas may also be linked with phrases like 'leads to', 'results in' and 'impacts on'. The concept map in Figure 0.6 looks at changes of state of matter and what causes these changes.

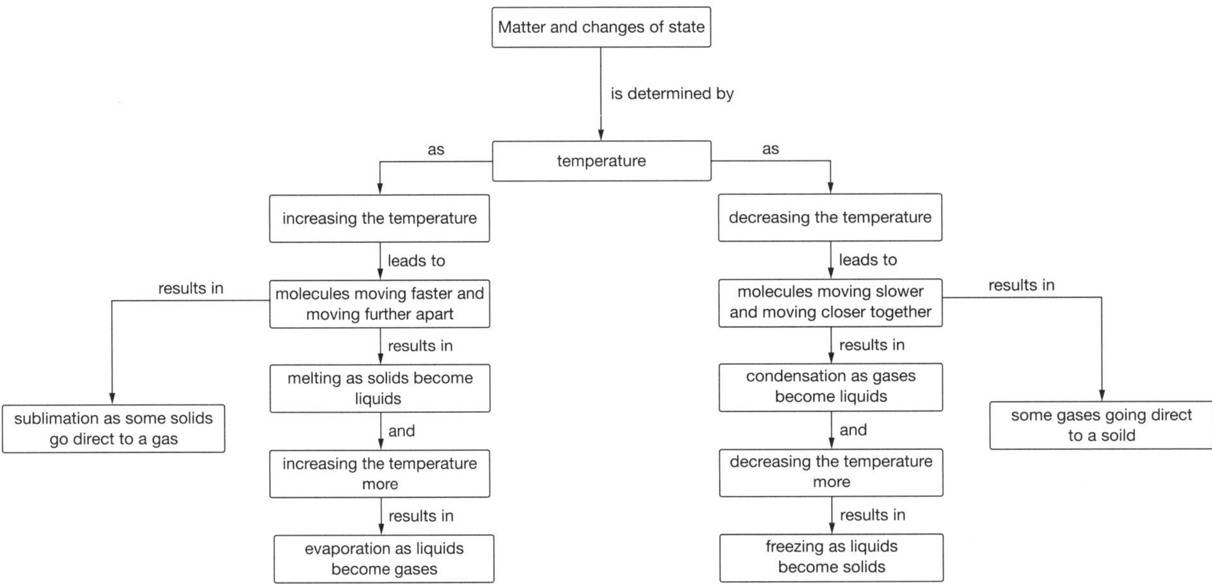

Figure 0.6 Concept map of the changes of state

Working with data

Scientists use lots of information or data. Data often comes in the form of numbers or figures and in science this is known as **quantitative** information. Examples of quantitative information include changes in temperature, the height and width of an object or how much liquid an object can hold. Data can also be qualitative. Data can be presented in many different ways.

Tables

Tables are a clear and simple way to record information. Often it is easier to read and understand data when it is presented in a table rather than a paragraph of text. The table must have a title that identifies information presented and labels for the rows and columns of data. Tables can be quite complex and contain lots of data. You need to be able to read a table accurately in order to understand the data presented. Table 0.2 shows a set of data from an experiment investigating the amount of gas produced from a reaction of a magnesium strip in acid. One conclusion we can draw from this table of data is that after 150 seconds, there was no further increase in the amount of gas collected every 30 seconds.

Volume of gas collected when a magnesium strip reacted with acid	
Time of reaction (s)	Volume of gas bubbles collected (mL)
0	0
30	29
60	57
90	82
120	92
150	95
180	95
210	95
240	95
270	95
300	95

Table 0.2 Table demonstrating a set of data from an experiment

Graphs

A graph is a diagram that shows the relationship between two variables. For example, a graph might show how the heart rate of a person increases as they run faster or it could show how many animals are endangered in the states and territories in Australia. Two of the more common types of graphs are column/bar graphs and line graphs.

Column and bar graphs

Column and bar graphs both display data using bars with the length of the bar indicating the data value. The difference between column and bar graphs is their direction: bar graphs are drawn with horizontal bars while column graphs have vertical bars.

The table of data below was used to draw a column graph (Figure 0.7). Steps for drawing the graph are listed below.

Average time spent on homework during the week for secondary school students	
Year Level	Average time spent (in hours) on homework during the week (Monday–Friday)
Year 7	2.5
Year 8	3
Year 9	4.5
Year 10	5
Year 11	6.5
Year 12	8.5

1 Draw the two axes. The x-axis runs horizontally and will represent the year levels of the students. The y-axis runs vertically and will represent the hours spent on homework. Grid paper will help you to accurately draw your graph; however if you don't have any, make sure you use a ruler to carefully draw your axes.

2 Label the axes using the same labels the table uses.

3 Work out the scale. Look at the largest number, which in this case is 8.5 (hours). The graph will need to go up to at least 9 hours. Because the graph also needs to show 2, 3, 4, 5 and 6 hours too, it makes sense to go up in 1 hour spaces or increments. The scale could therefore be 1 cm = 1 hour.

4 Draw in the bars. Do one year level at a time.

5 Add a title to the graph.

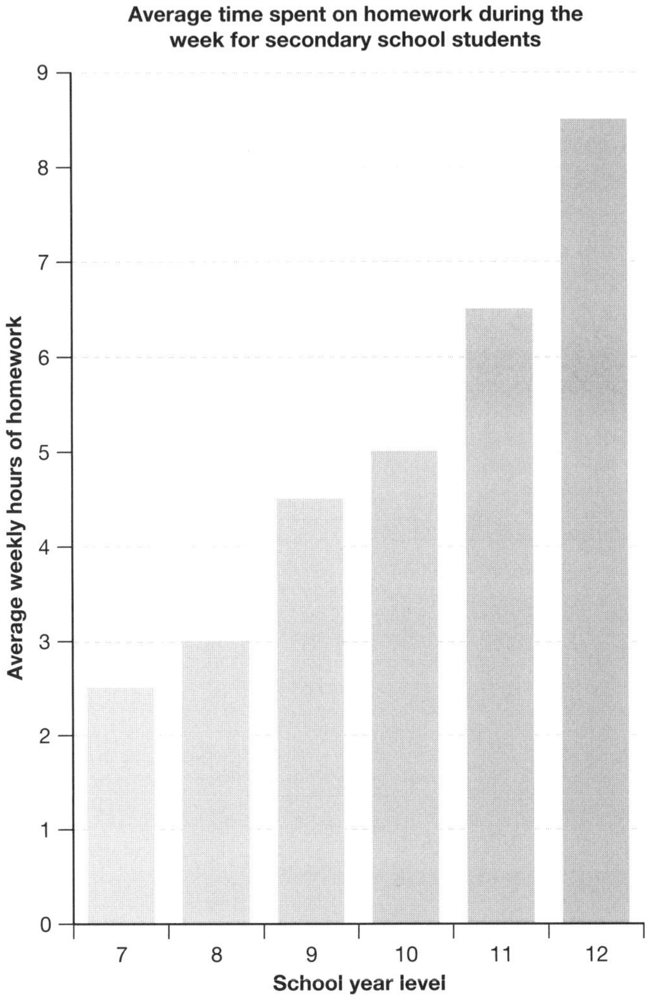

Average time spent on homework during the week for secondary school students

Average weekly hours of homework

0 1 2 3 4 5 6 7 8 9

7 8 9 10 11 12

School year level

Figure 0.7 Column graph of time spent on homework at different year levels

Line graphs

A line graph plots data as a point on the graph and then all the data points are joined together to form a line. A line graph is often used when both sets of data are numerical. Just like bar graphs, the independent variable (the thing that can be manipulated) is put onto the x-axis while the dependent variable is drawn along the y-axis.

A line graph (Figure 0.8) is drawn using the data below. Data shows the average maximum heart rate achieved while exercising for different age groups.

Exercise and heart rate	
Age (years)	**Average maximum heart rate (beats per minute)**
20	200
30	190
40	180
50	170
60	160
70	150

To draw this graph, follow the same general steps as for the column graph, plotting points and a line instead of columns.

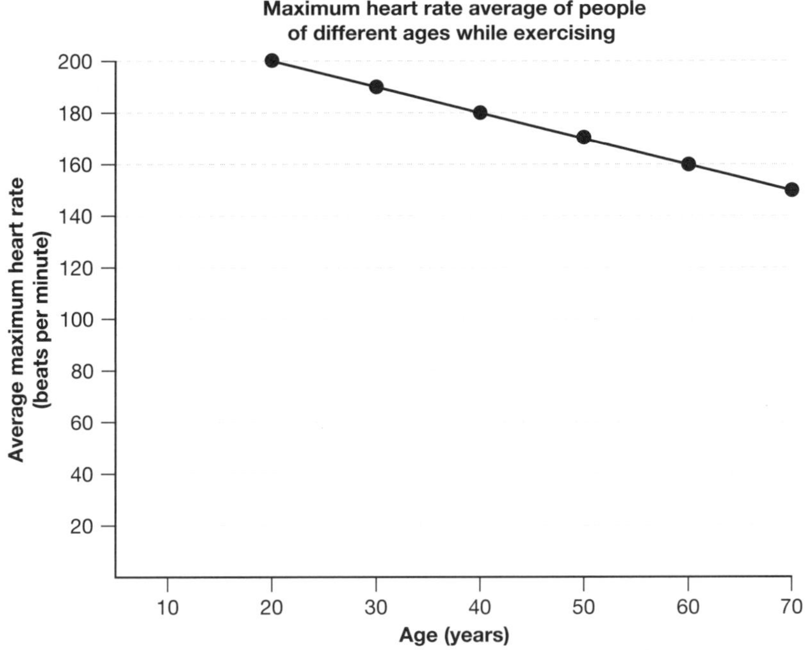

Figure 0.8 Line graph of heart rate while exercising, by age group

SALTS: A checklist for drawing graphs

Use the following checklist when drawing graphs:

- Scale: Have I used a suitable scale for the range of data?
- Axes: Have I ruled my axes?
- Labels: Have I labelled the two axes?
- Title: Have I given the graph a title?
- Source: Have I provided a source for the data (where necessary)?

Scientific investigations

One of the roles of a scientist is to make observations and conduct experiments in order to find out the answers to their questions. This is known as the scientific method. The results from these scientific investigations are then written up in investigation reports.

The scientific method helps scientists report on what they did, what happened and how their findings contribute to a better understanding of the subject. The reports you complete in class should aim to do the same thing. A report is not a story. A good report includes a purpose, aim, hypothesis, materials list, safety notes, procedure, results, discussion of the results and procedure, and a conclusion analysing and summarising the findings.

Figure 0.9 on pages xiv and xv is an example of a report for an investigation of growing cress for food in a space environment. Explanations of how to complete the criteria for each section of the report are provided. The results and discussions sections are samples only.

Science Toolkit

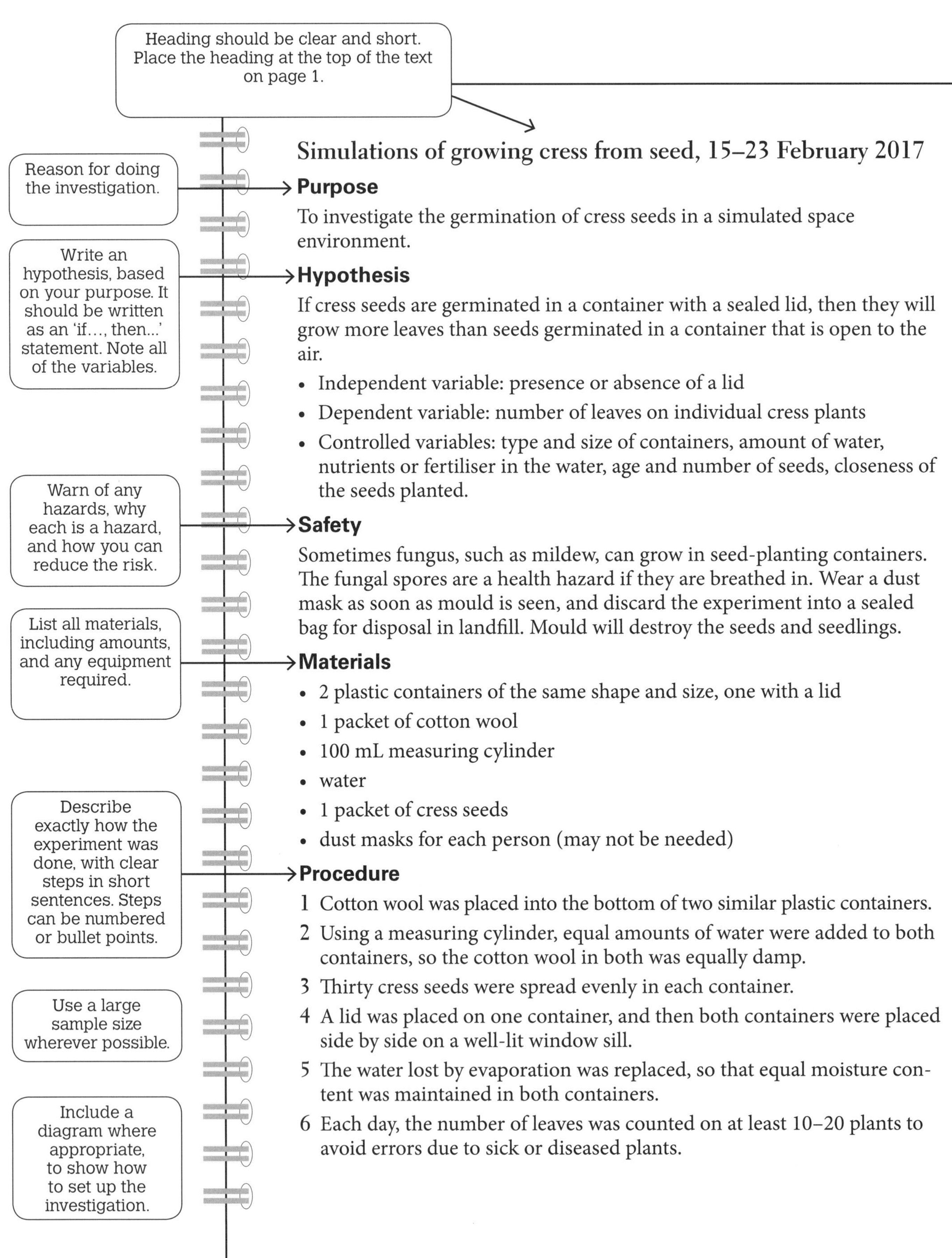

Record your results in a table, using the accepted format. As you observe your experiment, record the data. Data may or may not change during the time of the experiment. Examples of data are pH values, temperatures, measurements of growth, colour, distance and so on. Data can also be shown in graphs, sketches and photographs.

Results

A sample table (only part shown):

Time (days)	Number of leaves on each plant			
	Covered with lid		No cover or lid	
	total number of leaves	Average	total number of leaves	Average
1				
2				
3				

If you take several readings of the dependent variable, you can calculate the mean (average). Then your results will be more reliable.

Discussion

A sample discussion (only part shown):

The results showed that, after 7 days, cress plants in the covered container had slightly more leaves than plants in the uncovered tray. The numerical results were an average of 4.6 leaves per plant when covered, compared to 3.7 leaves per plant when uncovered. The reason for this could be due to….

This is the meaning or interpretation of results.
- What do your results show?
- Is there some reason why the results are not as expected?
- Were there any problems during the experiment? What did you do to overcome them?
- Do the results support the hypothesis?

Conclusion

A sample conclusion (only part shown):

Cress seeds grown in a sealed container had 24% more leaves than cress seeds grown in an open container. This supports the hypothesis that… because…

This relates back to the purpose. The conclusion should be clear and short. Someone wishing to learn more about experiments like this will read the purpose and the conclusion, to see if the investigation can help them in their research.

Avoid statements such as 'always' or 'never' in science writing. Refer to what the data shows—all claims must be supported by evidence.

Figure 0.9 A model investigation report

Research skills

Time management

Before starting a research task, plan how you will manage your time effectively so that the work is spread out and you don't leave everything until the last minute. Research can be time-consuming. It is easy to spend most of your time finding information and then rush the remainder of the steps of your project. It is important to plan your time for each stage of the project. Look at the suggested planning timeline in Figure 0.10.

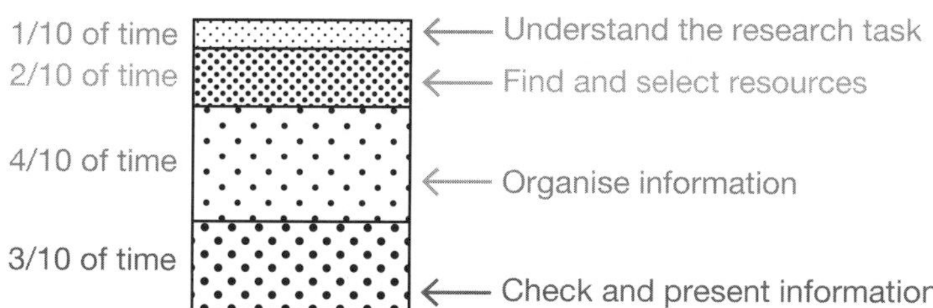

Figure 0.10 Time planner for research tasks

Tips to help you manage your time effectively:

- Block out periods of time after school and on the weekend and dedicate this time to completing your research task.
- Find a clean, comfortable space where you can complete your task and where you will not be constantly disturbed or distracted.
- Decide what part/s of the task will take the longest and consider completing them first.

If you are working in groups, make sure that the workload is spread evenly and that everyone knows what they are responsible for. A responsibility chart like the one below may help you to make sure everyone is on the right track.

	Person responsible	Date to be completed	Task completed? Y/N
Tasks			

The research question

The first step in completing a research task is to understand exactly what you are expected to do. Look at the key words in the question or task. Underline or highlight the key words and any words whose meaning you do not know. Look up the meanings of these words. Then rewrite the question or task in your own words. See the example in Table 0.3.

Research task	Humans have two kidneys but are able to live with only one. Evaluate whether you would donate a kidney. Justify your choice.
Rephrased research task in student's own words	Explain what the function of kidneys is. What are the negatives and positives of donating a kidney—for donor, for recipient? State if I would donate a kidney and back it up with reasons.

Table 0.3 Example of a research task and how it can be rephrased for better understanding

Some common terms used in research tasks and their meanings are listed.

Key terms:

Discuss—examine the main ideas, explore reasons for and against and draw conclusions

Explain—describe how something works or how it came to be; break things down into smaller parts

Identify—explain the key parts of something

Compare—discuss the features or qualities that two or more things have in common or that differ between two or more things

Contrast—discuss the differences between two or more things

Evaluate—provide your own opinion on the worth/value of something; explore both positive and negative aspects before providing your evaluation

Investigate—look into something

Analyse—take apart an idea, question or statement in order to look at all parts of it

Predict—suggest what might happen based on the information you have

Construct—make, build or create something

Simple questions to guide research

Sometimes research tasks may be written by your teacher with specific instructions and questions on a topic, or they may be open ended. Your teacher may select a broad issue or theme and ask you to choose a topic and develop your own research questions. For example, the broad theme might be 'the environment' and you may decide to research 'water conservation' or 'waste management'.

To guide your research, follow these six steps:

1 Identify the purpose of the research.

2 Understand the assessment criteria.

3 Write questions to guide your research.

4 Understand and use suitable key words in the questions.

5 Plan and manage your time to complete the task by the due date.

6 Consider how your research will be presented.

One way to approach the writing of research questions is to start with Who, What, Where, When, Why and How. This helps you to cover all aspects of a topic.

If you choose to research water conservation, the research questions could include:

1 Who is able to help conserve water?

2 What are the most effective ways of conserving water in the home?

3 Where in Australia is the most water wasted?

4 When is the best time of the day to water your garden?

5 Why do we sometimes have water restrictions?

6 How can we conserve water at school?

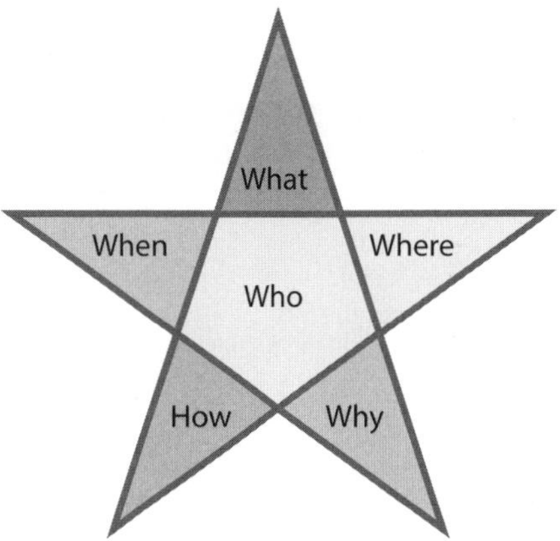

Figure 0.11 Who, what, when, where, why and how star

Finding resources

When looking for resources, you should try to draw from a wide variety of places so that you are able to choose a broad range of resources (Figure 0.12). You need to focus on picking the best resources to suit the task.

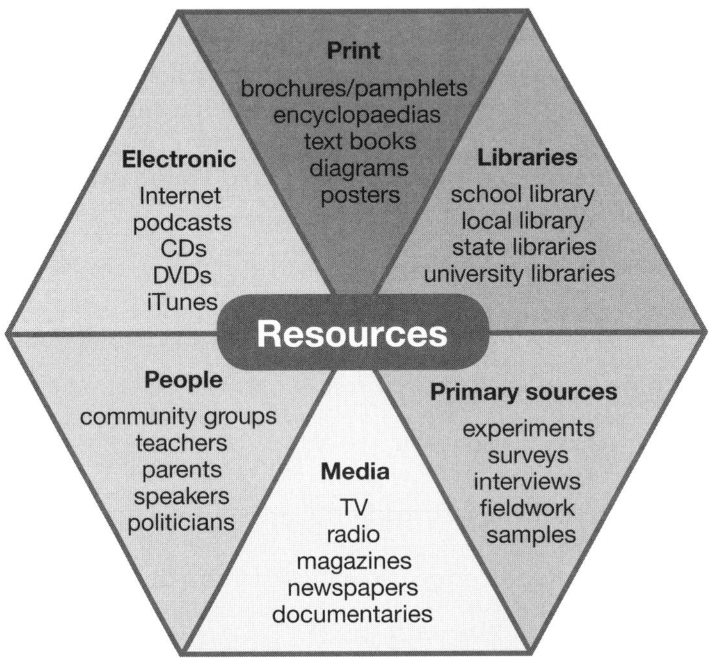

Figure 0.12

Using a wide range of resources is important; however, it's also worth noting that every type of resource has its advantages as well as disadvantages (Table 0.4). The table below lists the advantages and disadvantages of the main sources of information.

Table 0.4 Sources of information

Type of resource	Advantages	Disadvantages
books	– information is easy to find through the table of contents and index – has been proofread and edited by an editor – usually well researched and written by an expert – easily accessible	– information can be outdated – can take time to locate the exact information you require
internet	– access to a huge amount of information on the one topic – the information can be updated easily	– anyone can publish work on the Internet—can be difficult to work out what information is accurate and reliable – can take time to locate exact information you require
magazines/journals	– are usually up to date and contain recently researched information – easy to find information on specific topics	– can be very technical and difficult to understand – can be biased
encyclopaedias	– information is usually clearly presented, easy to follow and easy to find – reliable as checked by an editor	– older versions may be outdated – may not be as detailed as some other resources
films/documentaries	– can provide first-hand accounts of events/situations – engaging, easy to watch	– difficult and time-consuming to go back and find specific information – may not be accurate and can be biased

Choosing the best resources

Scanning and skimming help you quickly look over a number of resources so you can decide whether they are useful or not for the research.

Scan by quickly looking over the resource. Check the title, subtitle, contents list and index for key terms. Check things like maps and diagrams. Consider whether your topic is covered by this resource.

Reduce your list further by skimming over the resources. Skim read the introduction and first paragraphs. Skim read the topic sentences (first sentences) of a few paragraphs. Then decide whether the resource will be useful for research. This will result in a manageable number of resources for you to use.

Presenting the research

Choosing the presentation format

There are many different ways to present your research task. Sometimes your teacher may direct you towards how you need to present your research and other times, you may choose the format yourself. Consider your audience and the purpose of your research task when thinking about how to present your work. Table 0.5 lists the characteristics of different presentation formats.

Table 0.5 Presentation formats

Format	Characteristics	Things you should include or remember
Poster	• a visual display of information • a summary of ideas • suitable for presenting information to many people	• a title that attracts attention • large headings that stand out • subheadings of a smaller size • attractive presentation • a balance of written and visual material • writing large enough to read from a distance
Pamphlet	• a small folded booklet with facts and diagrams • a summary of ideas • an easy way to access information • suitable for providing take-home information to individuals	• consistent layouts and fonts • dot points (if appropriate) • diagrams • attractive presentation
Report	• a presentation of clear and detailed information on a topic • suitable for presenting detailed information to individuals	• an introduction, paragraphs and a conclusion • subheadings • mainly text, but can include diagrams, maps, tables and/or graphs
Retrieval chart	• very commonly used • very useful for summaries of notes • very useful for organising statistics or data • a way of collecting information that can be presented in another format • a table of information • suitable for presenting to a large audience or an individual	• information in sentences or point form

Format	Characteristics	Things you should include or remember
Microsoft PowerPoint slide show	• an easy-to-follow format • a good format for presenting to a large audience • a useful tool if used with an oral presentation • able to provide a summary of information covered in an oral presentation • able to be printed as a set of notes to be given to the audience	• visual and written information • consistent backgrounds, formats and colours throughout • minimise the amount of text on any one slide
Oral presentation	• able to inform and possibly persuade the audience • engaging and can be entertaining • a good format for presenting a point of view on an issue • suitable for presenting to a large audience	• use relevant anecdotes • use palm cards to glance at if needed • rehearse your speech so you don't read the palm cards word for word • keep eye contact with the audience • speak clearly and at a volume that can be heard • don't fidget or wriggle around • stand up straight and look confident
Model	• a three-dimensional visual display • a good way to provide information with very little text • suitable for communicating information to a lot of people	• a title, labels and explanations
Website	• able to present visual and written information • accessible to a worldwide audience • easy to follow	• hyperlinks to related information • multimedia, such as video clips and audio (if appropriate) • consistent backgrounds, formats and colours throughout • headings that stand out • list all of the site hyperlinks on the main page • the author's name and credentials, and the date of publication

CHAPTER 1

Working with scientific data

1.1 Knowledge preview

Science understanding

FOUNDATION	STANDARD	ADVANCED

How well do you understand and work with scientific data?

Complete the activity below to gain an idea of your prior knowledge. When you complete the 'Working with scientific data' chapter, look at this activity again. What would you change? Scientific investigation is an extremely important part of working scientifically. The report written about a scientific investigation has sections and follows a particular order.

- The flow chart on the left-hand shows the sections of an investigation report. Two section names have been filled in as an example.
- Connected to each investigation report section is a box to list what is included in that part of the report. For example, the section on materials might include items such as a beaker, filter paper etc. as shown below.
- On the far right of the chart is a jumbled list of items to include in a report.

1 Fill in the missing sections on the left column of the chart. Decide which section of the chart each item belongs to.

- write the item into the correct section of the chart
- use each term once only
- there may be more than one item entered into the 'Section details' boxes

Scientific investigations report chart

Investigation report sections	Section details
heading	
materials	beaker, filter paper

List of items to be written into chart

- the method used
- comparison with hypothesis
- data
- procedure
- beaker
- the aim
- investigation name
- materials
- discussion
- 1 g sodium chloride
- mix 1 g sodium chloride in 1 mL of water
- summary of findings
- describe findings
- variables
- results
- line of best fit
- purpose
- bar graph
- filter paper
- educated guess of result
- heading
- hypothesis
- conclusion

1.2 Spot the difference

Science inquiry skills

FOUNDATION | **STANDARD** | ADVANCED

Processing & Analysing | Questioning & Predicting

Scientists need to be able to pick out the similarities and differences in what they see around them. These similarities and differences allow them to classify all sorts of things such as rocks, stars, chemicals and living organisms.

Compare the following pairs of organisms by using Venn diagrams to show their similarities and differences. In comparing the items, think about characteristics such as number of legs, antennae, leaflets, shape of body, beak, horns, veins in the leaves, and so on.

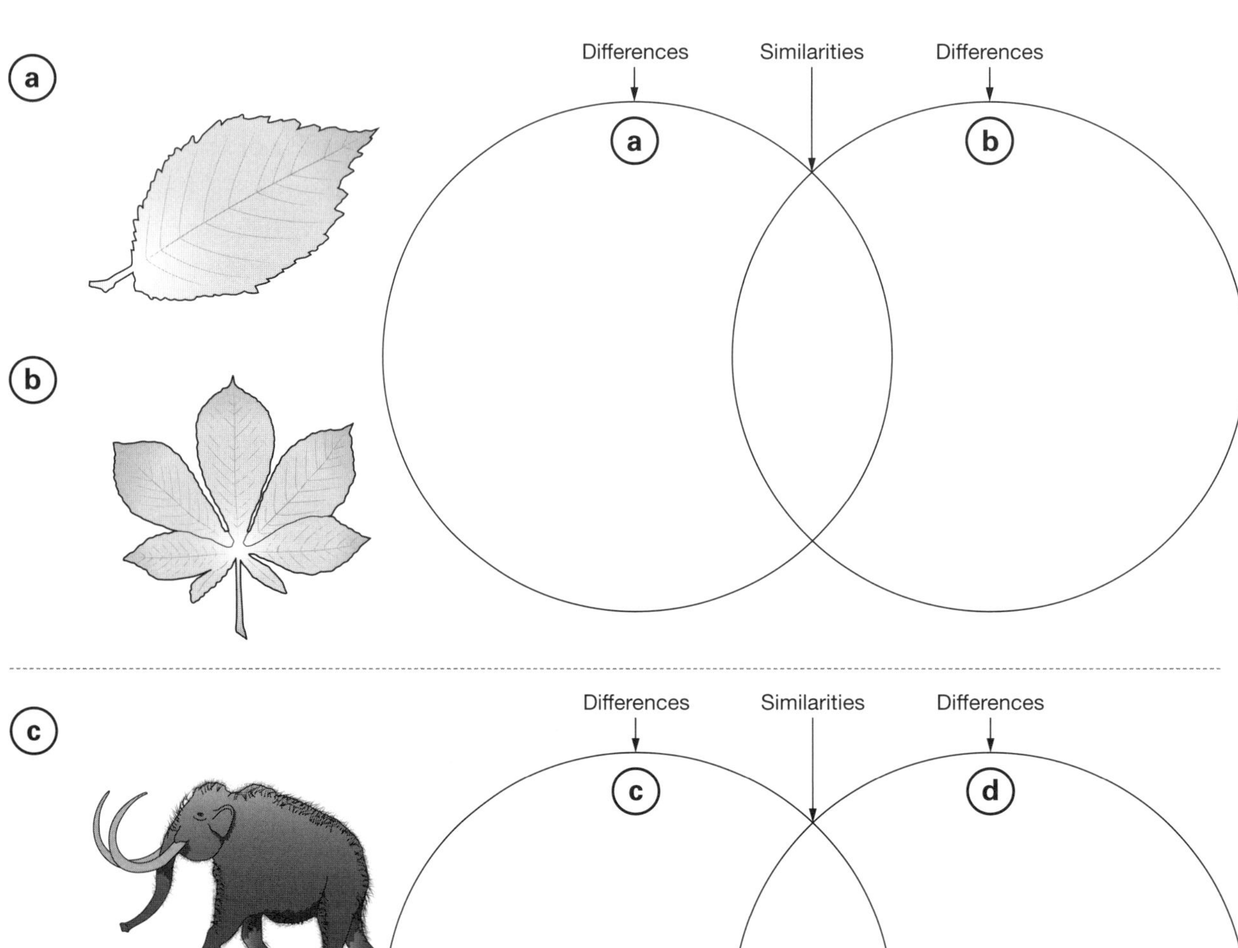

d

1.2 Spot the difference

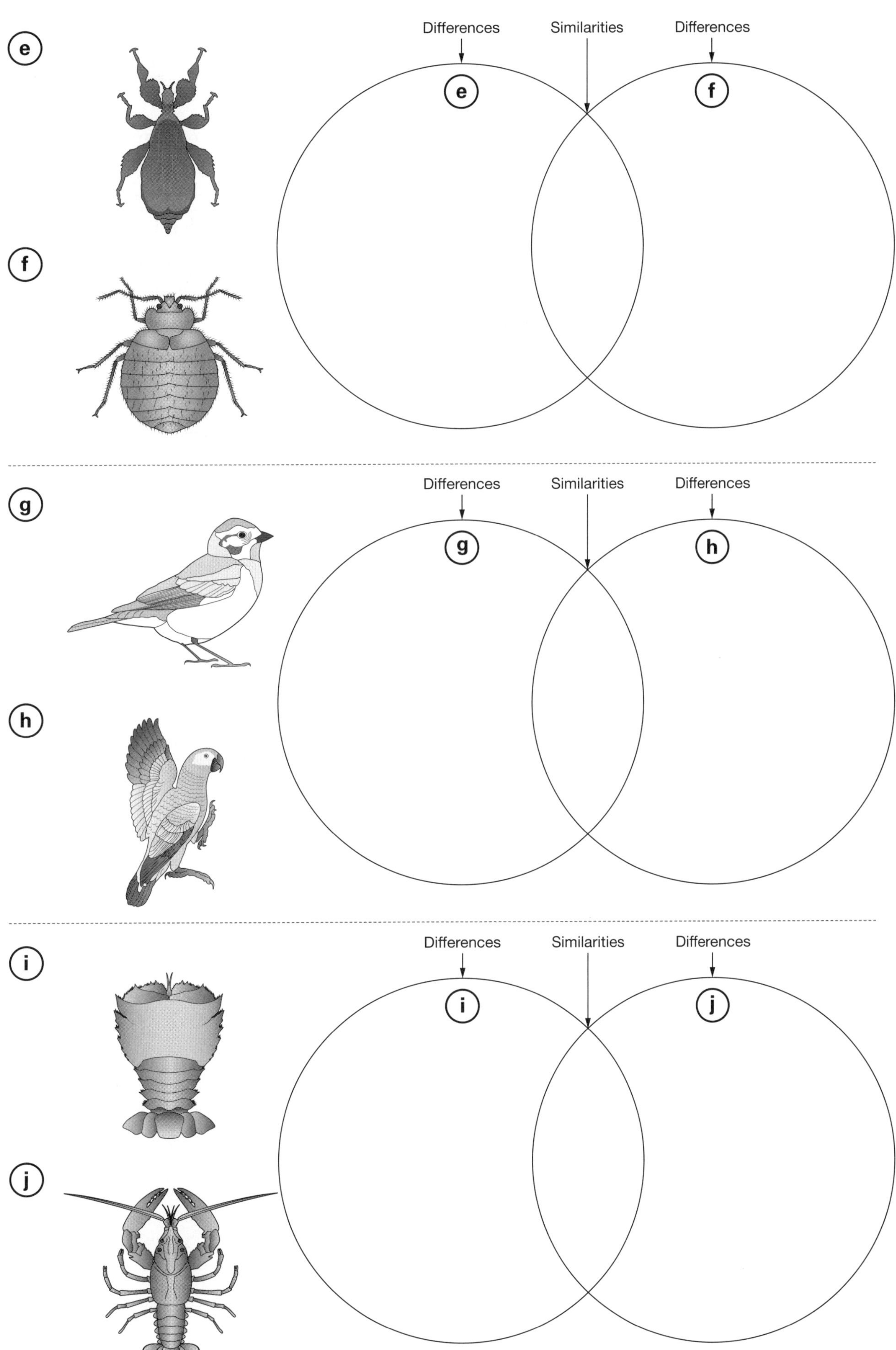

RATE MY UNDERSTANDING
Shade the face that shows your rating

1.3 Patterns in observation

Science inquiry skills

FOUNDATION | **STANDARD** | ADVANCED

Questioning & Predicting

Scientists need to observe the world around them carefully and be able to recognise patterns in what they see. In each of the following, use the patterns to predict the final missing design.

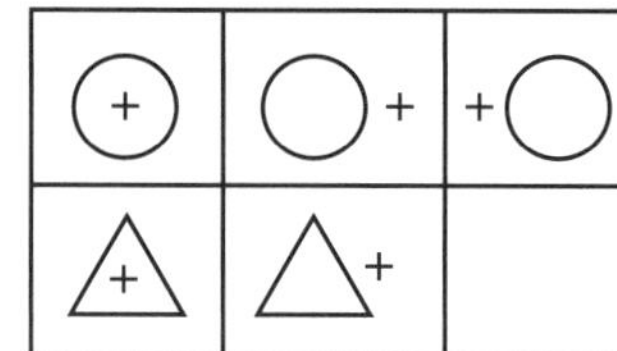

b

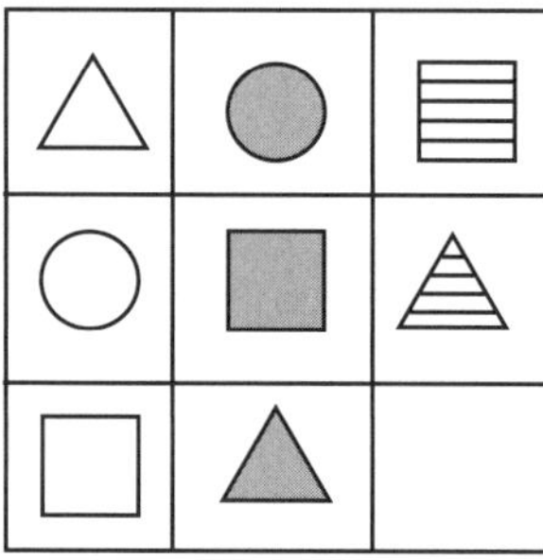

c

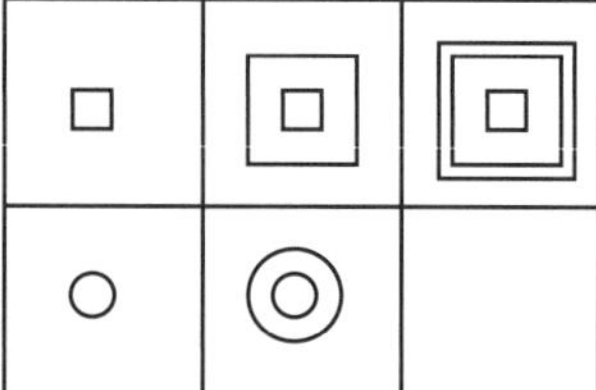

d

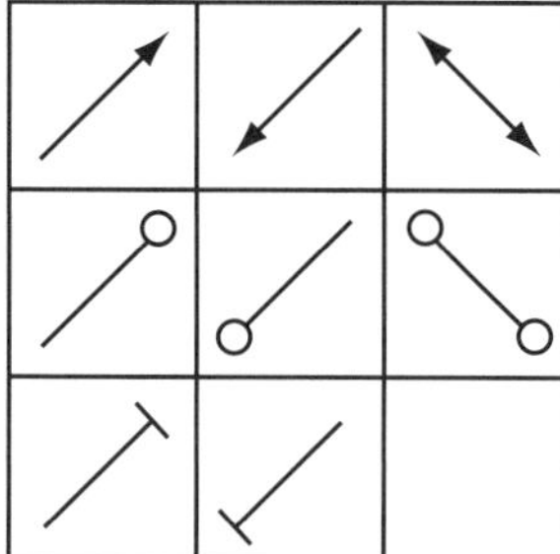

e

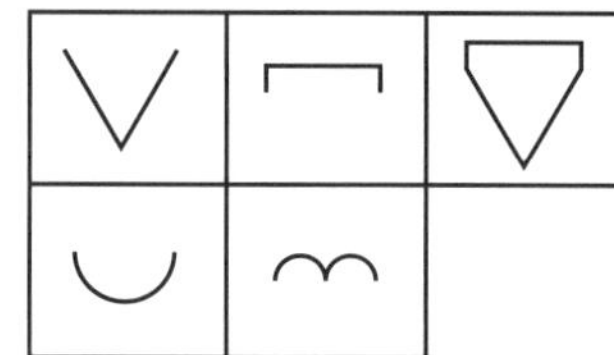

f

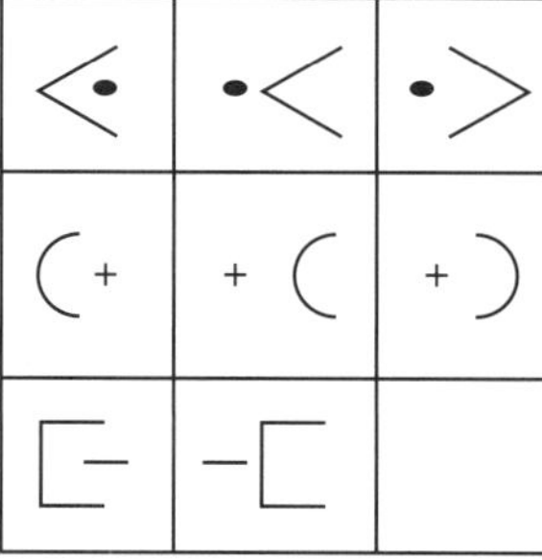

g

h

i

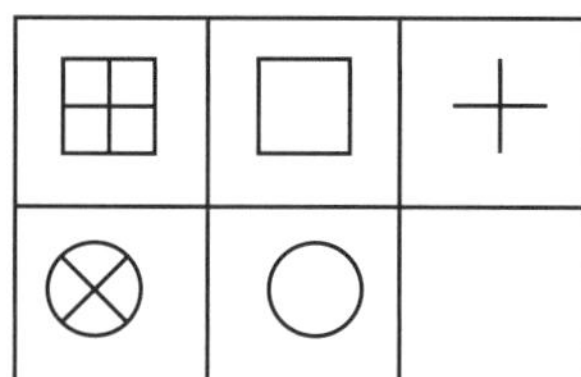

j

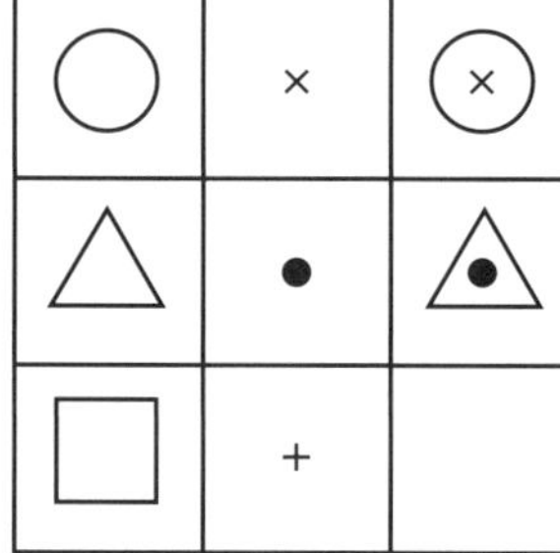

RATE MY UNDERSTANDING
Shade the face that shows your rating

1.4 Equipment

Science inquiry skills

FOUNDATION	STANDARD	ADVANCED

Planning & Conducting

Scientists rarely draw realistic three-dimensional (3D) diagrams of the equipment they use. Instead, they draw simple two-dimensional (2D) diagrams showing each piece of equipment as a cross-section, as shown in the diagram below.

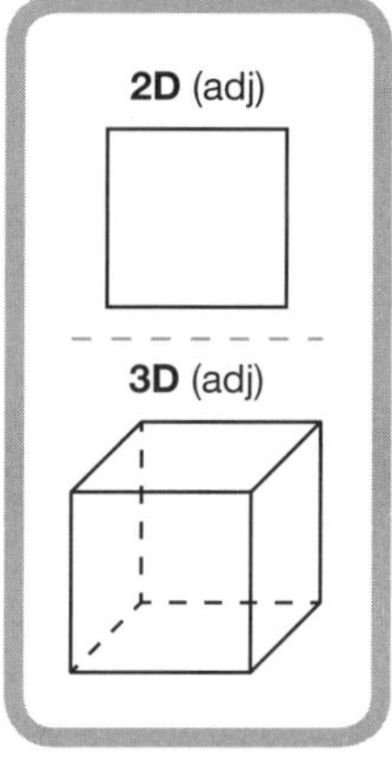

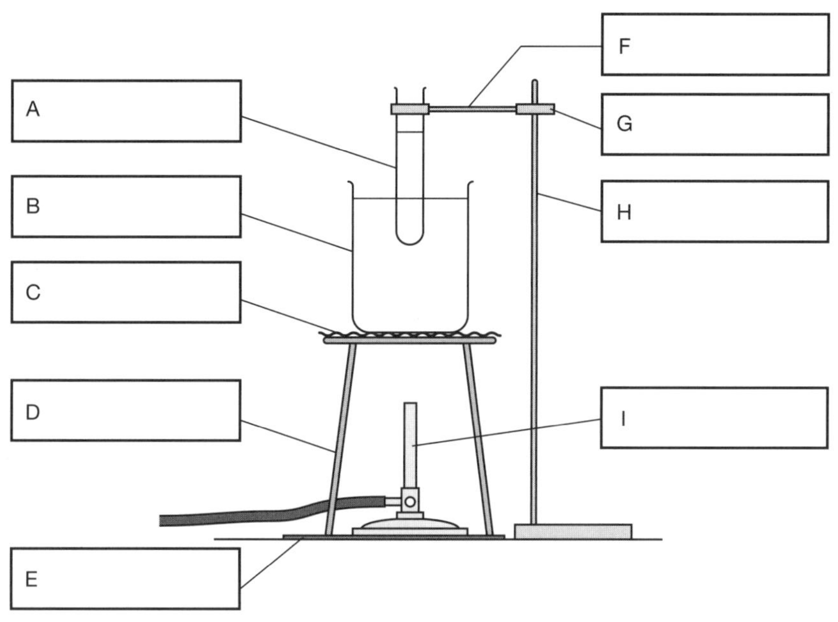

tripod
Bunsen burner
beaker
conical flask
test-tube
filter paper
retort stand
bosshead
funnel
clamp
bench mat
gauze mat

1 Use words from the box to label all the pieces of equipment shown above.

2 Suggest what this set-up of equipment might be used for.

____

____

3 Here is another experiment, this time drawn in 3D. Use words from the box to label all the pieces of equipment. Then draw a 2D diagram of this set up in the space provided.

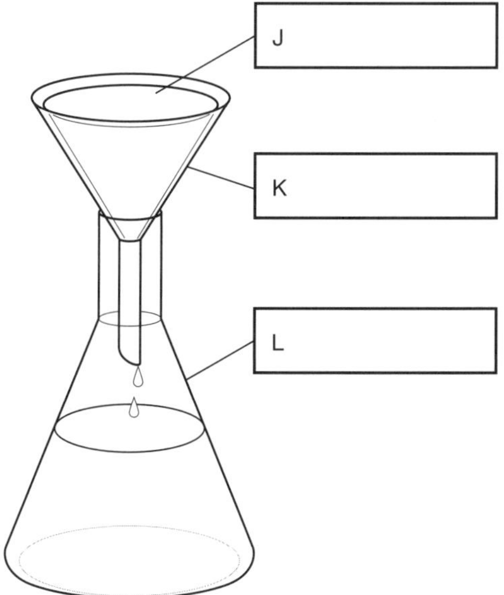

RATE MY UNDERSTANDING
Shade the face that shows your rating

1.5 Can you follow instructions?

Science understanding

FOUNDATION	STANDARD	ADVANCED

To complete this activity you need to follow the instructions given below as quickly as you can.

Instructions

1. Don't do anything until you have carefully read every instruction.
2. Write your full name in capitals. ______________________
3. Circle your surname.
4. Name the capital city of Australia. ______________________
5. Fold over the top corner of this page.
6. State how many kangaroos are on the $1 coin. ______________________
7. Complete the following diagram to construct a hexagon (six sides).
8. Specify which Australian coin has a platypus on it.

   ____
9. Draw a smiley face in the bottom left corner of this page.
10. Name Australia's current prime minister. ______________________
11. Calculate the answer to 345 + 289 + 365. ______________________
12. State today's date. ______________________
13. Name the value of the Australian banknote that is red in colour. ______________________
14. Write the letter E back-to-front, so that it appears as if it is reflected in a mirror.
15. Arrange the following numbers in order from smallest to largest: 21, 8, 64, 19, 3.

    ____
16. Draw a square in the bottom right corner of this page.
17. Write the number 8 inside a square.
18. Ignore all the above instructions except instructions 1, 2 and this one!
19. What did you learn by following these instructions?

    ____

    ____

RATE MY UNDERSTANDING
Shade the face that shows your rating

1.6 Mega quantities

Science inquiry skills

FOUNDATION | **STANDARD** | ADVANCED

Questioning & Predicting | Planning & Conducting | Processing & Analysing

Scientists use metric units such as metre (m), gram (g) and litre (L) to measure quantities. Sometimes it is inconvenient to use these units to measure extremely large quantities. This is when unit prefixes such as kilo (kg), mega (M) and giga (G) are used. On the other hand, when measuring really small quantities units such as centi (c), milli (m), micro (µ) and nano (n) are used.

> **gram** (symbol: g) a unit of mass
>
> **litre** (symbol: L) a unit of volume
>
> **metre** (symbol: m) a unit of measure

Mega (symbol M) is the prefix for a million. For example, 1 megalitre (ML) of water is the same as 1 million litres and 1 megatonne (Mt) represents 1 million tonnes. The prefix mega can be attached to anything. For example, 1 megapeople would represent 1 million people and so Australia has a population of around 20 megapeople. Likewise, you would be rich if you won 1 megadollar ($1 million dollars).

The following two tasks will give you some idea of how big a million or a mega is.

Task 1: A million letters

1 For this task, you need a couple of thick books such as novels. Estimate how many pages you think would be equivalent to 1 000 000 letters.

Estimate = ______ pages.

2 Once you've made your estimate, turn to a page in the book that is all text with no pictures. Don't choose the first or last page of a chapter because they usually aren't a full page. Count the number of letters in the first, second and third lines on the page. Record your results for questions 2, 3, 4 and 5 in Table 1.6.1 on page 8.

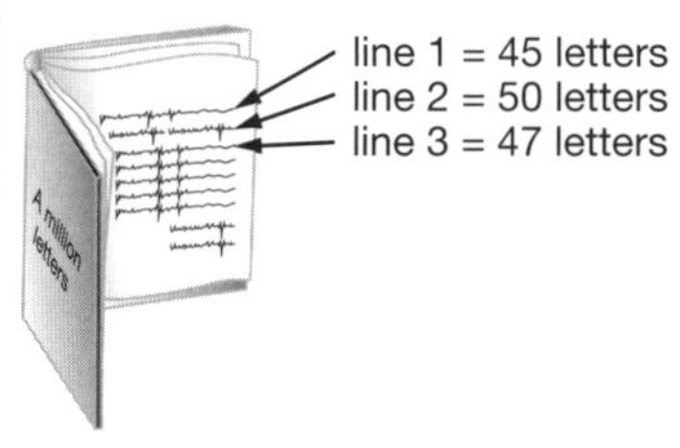

3 Calculate the average number of letters in one line by adding your three numbers and dividing by 3. Record your average in the table.

$$\text{average letters per page} = \frac{\text{line 1} + \text{line 2} + \text{line 3}}{3}$$

Worked example:

$$\text{Average letters per line} = \frac{45 + 50 + 47}{3} = \frac{142}{3} = 47.3$$

4 Count how many lines there are on one page and then calculate the average number of letters on one page. Record your average in the table.

$$\text{average letters per page} = \text{average letters per line} \times \text{number of lines on one page}$$

Worked example:

34 lines on one page

$$\text{Average letters per page} = 47.3 \times 34 = 1608.2$$

5 Calculate how many pages would be equivalent to 1 000 000 letters by dividing 1 000 000 by the average number of letters on one page. Record the number in the table.

$$\text{Number of pages} = \frac{1\,000\,000}{\text{average per page}}$$

Worked example:

$$\text{Number of pages} = \frac{1\,000\,000}{1608.2} = 621.8 \text{ pages}$$

Copy all your results into this table.

A million letters results table	
letters on line 1	
letters on line 2	
letters on line 3	
average letters in 1 line	
lines on 1 page	
average number of letters on 1 page	
number of pages that make up 1 000 000 letters	

6. Compare your estimate in question 1 with the number of pages you calculated in question 5.

____

Task 2: A million heartbeats

7. Estimate how long it would take for your heart to beat 1 000 000 times.
Estimate = ______________ minutes, hours or days

8. Construct a method showing how you could calculate how many days it would take for your heart to beat 1 000 000 times (Table 1.6.2 in question 9 might give you some ideas how this could be done). Remember there are 60 minutes in an hour and 24 hours in a day. Write your method below.

____

____

____

____

9. Carry out your method and record all the results in the table.

A million heartbeats results table	
average beats per minute	
beats per hour	
beats per day	
days taken for 1 000 000 beats	

10. Compare your estimate with the number of days calculated in the table above.

____

RATE MY UNDERSTANDING
Shade the face that shows your rating   

1.7 Analysing graphs

Science inquiry skills

FOUNDATION | **STANDARD** | ADVANCED

Processing & Analysing

Friction

Friction is the force that causes your bike to slow down and is the force that stops a car at the traffic lights. Friction depends on the type of tyres and the type of surface being travelled on. Some surfaces are extremely smooth and have low friction while others are very rough and have a lot of friction. The coefficient of friction measures how rough a surface is. This bar graph shows the coefficients of friction for different types of road surfaces. The higher the coefficient, the more friction the surface provides.

coefficient (*n*) a number that measures one thing about a substance

rank (*v*) to place in order from lowest to highest or highest to lowest

skid (*n*) the slipping or sliding of something such as car wheels

Coefficients of friction for different types of road surfaces

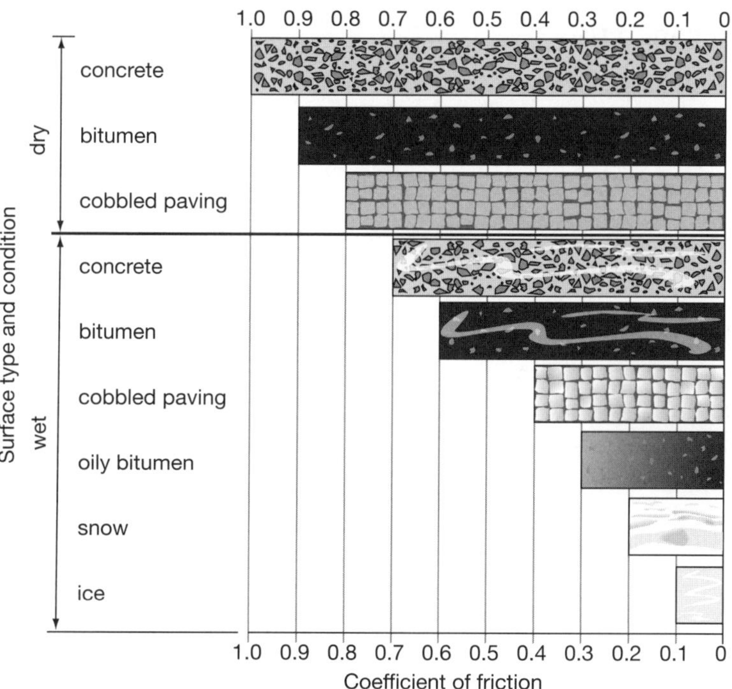

1 What is friction? ______

2 Use the bar graph to rank the different surfaces in order from the surface that provides the most friction to the surface that provides the least.

______

3 Identify the surface on which it would be:

(a) easiest to stop ______

(b) most difficult to stop. ______

4 Identify which surface would produce the biggest skids. Justify your choice.

______

5 Friction also allows you to get moving. Without it, your bike wheel would just spin on the spot. Identify the surface on which it would be:

(a) easiest to get moving ______

(b) most difficult to get moving. ______

6 Compare the coefficients of friction of the following road surfaces.

(a) dry bitumen, wet bitumen and oily bitumen

______

(b) dry concrete and ice

______

(c) dry concrete and wet concrete

______

(d) snow and ice

______

UV counts

Ultraviolet (UV) radiation from the Sun is known to cause skin cancers. The amount of UV radiation falling on Australia depends on the time of year and the time of day. This variation is shown as the UV count in the line graph.

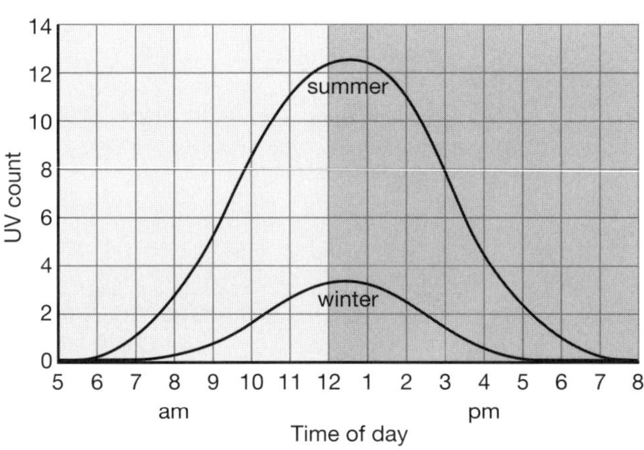

7 Use the line graph to state the maximum UV count for:

(a) summer

______

(b) winter.

______

8 UV counts of 6 or higher can be dangerous and a health risk. Specify the times of the day at which the UV count is 6 or higher in:

(a) summer ______ **(b)** winter ______

9 Propose a reason why the UV counts in January would be higher than those in July.

______

10 State the hours between which the UV count is zero in:

(a) summer ______ **(b)** winter ______

RATE MY UNDERSTANDING
Shade the face that shows your rating

1.8 Planning your own experiment

Science inquiry skills

FOUNDATION	STANDARD	ADVANCED

Planning & Conducting

Imagine that you have been given the task of finding what affects the time it takes for a parachute to drop.

1. List as many variables as you can think of that might affect the drop time of a parachute.

____

____

____

HINT

Variables are different possibilities.

A purpose is the reason for the experiment. Begin with: To ...

The hypothesis is the idea or explanation that is being tested.

2. Identify the variable that you think will have the biggest effect on a parachute's drop time.

____

3. Imagine that you now need to carry out an experiment that tests this variable. Before you start, write a:

(a) purpose for your experiment

____

(b) hypothesis for your experiment

____

____

(c) list of materials (including equipment) you will need

____

____

(d) step-by-step procedure of how you will carry out your experiment. (Remember that you should only change the variable you are testing.) Your procedure should give enough detail so that another Year 8 student would be able to repeat your experiment exactly the same way.

____

____

____

____

____

____

RATE MY UNDERSTANDING
Shade the face that shows your rating ☺ 😐 ☹

1.9 Investigating the pendulum

Science inquiry skills

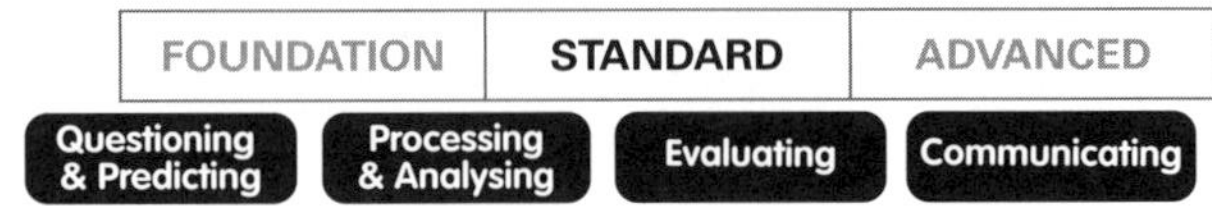

A pendulum is a mass on the end of a string, cable, chain or bar that repeatedly swings back and forth (Figure 1.9.1). Nathan was investigating what affects the swing of a pendulum. The results he obtained in one experiment are shown in Table 1.9.1.

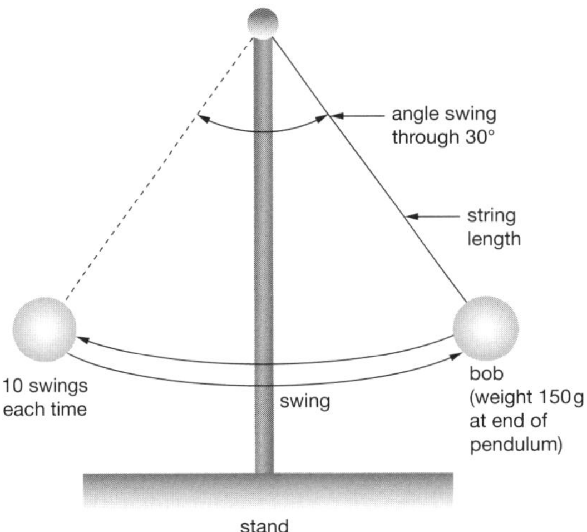

Figure 1.9.1 Pendulum swing

Mass (g)	String length (cm)	Angle swung from	Time for 10 swings (s)				Period = average time for one swing (s)
			Time 1	Time 2	Time 3	Average	
150	20	30°	9.12	8.95	9.08		
150	40	30°	12.53	13.12	12.74		
150	60	30°	16.30	16.03	15.25		
150	80	30°	18.41	18.84	18.38		
150	100	30°	19.89	20.11	20.87		

Table 1.9.1 Effect of string length on pendulum swing

1 From these results, identify :

(a) what Nathan altered in his experiment (independent variable)

____

(b) the two variables that Nathan kept the same throughout the experiment (controlled variables)

____

____

(c) the measurement that changed naturally as Nathan changed other variables in his experiment (dependent variable).

____

2 Calculate Nathan's missing values to complete his results table.

3 Suggest a reason why Nathan measured 10 swings each time and not just one swing.

4 Nathan measured each set of swings three times and then took an average. Explain the advantage of this.

5 On the graph axes below, construct a line graph showing period versus length for Nathan's experiment. Write a title for the graph. Include scales on each axis.

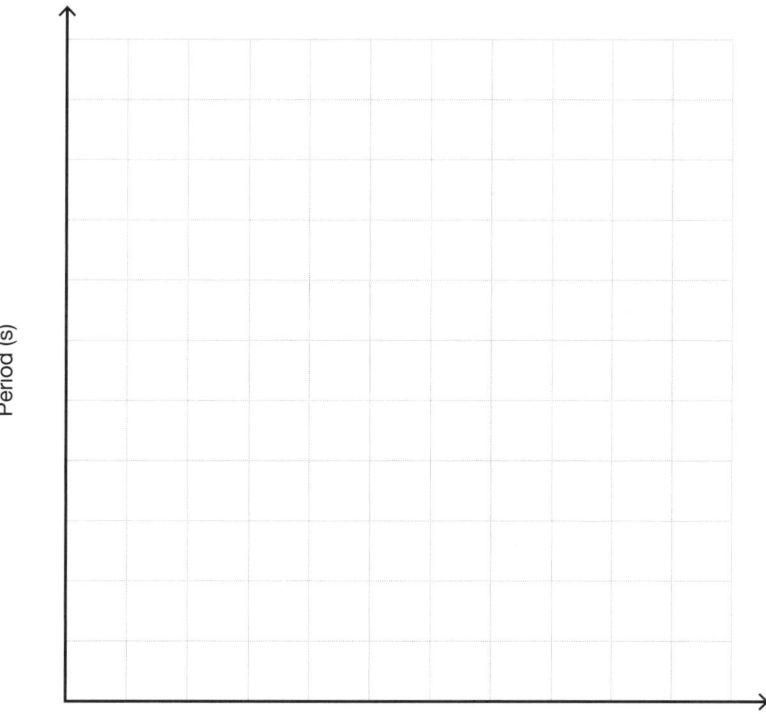

6 From the following sentences, identify the best conclusion to Nathan's experiment by highlighting the answer.

A The length of a pendulum does not affect the period of a pendulum.

B If the length of a pendulum doubles, then its period also doubles.

C As the length of the pendulum increases, its period increases at an increasing rate (that is, it keeps on increasing and increasing).

D As the length of the pendulum increases, its period increases but at a lesser rate than before (that is, it flattens out).

RATE MY UNDERSTANDING
Shade the face that shows your rating

1.10 Literacy review

Science understanding

FOUNDATION | **STANDARD** | ADVANCED

1 Match each term with its correct definition by drawing a line between them.

Definition
the mean; found by adding up all the values and dividing by the number of values
the method for an experiment
a type of error caused by looking at a measurement from a slight angle
an observation made using numbers
an observation that has no numbers but uses a description or diagram
a resource list is sometimes called this
the purpose of an experiment
logical extension of a graph
an educated guess about what might happen
a factor that might influence what happens in an experiment
factors that are kept constant
a short summary of what was found out in an experiment
data that you collect yourself in an experiment
altering an electronic scale so that it reads zero
often confused with errors; can be avoided with care
slight changes in measurements that occur naturally, regardless of how careful you are
measurements and observations about something
a source of second-hand data

Term
extrapolation
variable
quantitative
conclusion
aim
parallax
errors
internet
tare
hypothesis
bibliography
first hand
qualitative
average
mistakes
data
procedure
constants

(2) Rafael has completed an investigation looking into which light conditions seedlings grow the tallest. He placed paper towels on 3 petri dishes and added 5 mL of water to each. He carefully placed 20 grass seedlings each measuring 1 cm high onto each dish. Dishes were placed in different positions around the room in different light conditions: very light, in shadow and in a dark corner. Each dish had 5 mL of water on day 7 of the experiment. The temperature in the room was 15° Celsius throughout the 10 days of the experiment. Help Rafael by completing the following questions.

(a) Highlight the best statement to be used for the purpose of the investigation report.

A To keep all variables such as number of seedlings and dampness of paper towel the same.

B To have only one dependent variable to measure—height of grass seedlings.

C To see how different amounts of light impact seedling growth.

D To see how watering impacts seedling growth.

(b) Rafael has written some of the procedure for the report. Place a tick or cross next to each sentence to indicate whether or not it should be used in the report. Rewrite any sentence you rejected with a cross, so that it is suitable for use in the report.

Seedling growth experiment report procedure		
Rafael's sentence	**Tick/ Cross**	**Corrected sentence**
Use paper towel on the bottom of the petri dishes.		
One petri dish was placed in a dark, smelly corner.		
Three petri dishes were used and a paper towel placed on each.		
The experiment went on for a long time.		
Each dish was placed in a position so different amounts of light were available to the seedlings: dark, in shadow and in sunlight.		
Some seedlings were placed on each dish.		

(c) Rafael has written his results but needs to convert them into a table. Read his results then complete the table which Rafael started.

> The investigation was carried out over 10 days. One dish was in a dark cupboard, another on a shelf in a shadow and another on a sunny window sill. The dishes were checked on days 1, 4, 7 and 10 of the investigation. All the seedlings in all three dishes were an average of 1 cm in height on day 1. The seedlings in the dark corner showed no signs of growth over the 10 days. The seedlings in shadow averaged a height of 1.5 cm on the 4th day, compared with 4 cm for the seedling in the sunlight. By day 7 the seedlings in shadow averaged 2.5 cm and by day 10 they were 4 cm high. The seedlings on the sunlight averaged 6 cm on day 7 and 9.5 cm on day 10.

Seedling growth experiment table of results				
Amount of light	**Average height of seedlings (cm)**			
	Day 1	**Day 4**	**Day 7**	**Day 10**
			1	
shadow		1.5		

(d) Highlight the sentences that are correct and are based on the observations of the investigation. These are sentences that Rafael could use in the discussion section of his investigation report.

(i) The seedlings in the dark grew the tallest.

(ii) Over the 10 days, all the seedlings grew at the same rate.

(iii) Seedlings in shadow grew taller than seedlings in the light.

(iv) The more light available to the seedlings, the taller the seedlings grew.

(v) The seedlings in the light were tallest by day 4 but by day 7 the seedlings growing in the shadow were the tallest of all.

RATE MY UNDERSTANDING
Shade the face that shows your rating  

1.11 Thinking about my learning

Check what you have learnt and understood about working with scientific data by completing the chart. Each row in the chart has an illustration and space in which to write terms that are related to the illustration.

(a) Select the appropriate terms from the list to the right of the chart. You don't have to use all the terms. Terms may be used more than once.

(b) Use all the terms you have selected and write one or two sentences explaining what you learnt about scientific data.

Investigation results	
Number of candles	Rise in water levels (mm)
1	2
2	4
3	6
4	8
5	10
6	12
7	14
8	16

Terms: ____________ ____________ ____________

I learnt that... ____________

____

____

____

Taking measurements

Terms: ____________ ____________ ____________

I learnt that... ____________

____

____

____

Kidney dissection

Terms: ____________ ____________ ____________

I learnt that... ____________

____

____

____

Constructing graphs

Terms: ____________ ____________ ____________

I learnt that... ____________

____

____

____

- data
- data representation
- error
- experiment
- first-hand data
- graph
- making observations
- mistake
- parallax
- qualitative data
- quantitative data
- second-hand data
- URL

CHAPTER 2

Cells

2.1 Knowledge preview

Science understanding

FOUNDATION	STANDARD	ADVANCED

Check what you know about cells and complete the following:

1 Write True or False for the following statements:

Statement	True or False	Statement	True or False
Microscopes and magnifying glasses both enhance your sense of sight.		Cells are the building blocks of living things.	
Animals are unicellular organisms.		A lung is an example of an organ system.	
The human body is made up of just one type of cell.		The brain is composed of nerve tissue.	
Most cells are composed of organelles.		Fungi are not made up of cells.	
All plant cells can make their own food.		Bacterial cells are smaller than plant and animal cells.	

2 A scientist investigated three different specimens and sketched pictures of them after observing them under a microscope. The scientist also included the magnification for each specimen. State whether each specimen is a plant cell, animal cell or bacterial cell and give a reason for your choice.

Specimen	Type of cell	Reason
x 400		
x 1000		
x 100		

3 Select the best answer to complete each statement from the choices provided.

(a) A micrometre is:

- **i** larger than a millimetre
- **ii** smaller than an atom
- **iii** a used to measure the size of cells.

(b) The heart is:

- **i** a tissue made up of blood
- **ii** an organ
- **iii** a muscle tissue.

(c) When a cell divides:

- **i** the number of cells increases
- **ii** the number of cells decreases
- **iii** the organism dies.

(d) Objects viewed through a microscope appear:

- **i** smaller and upside down
- **ii** larger and upside down
- **iii** the same size but back-to-front.

(e) When considering organelles, we know that:

- **i** cells are made up of only one type of organelle
- **ii** organelles have specialised functions
- **iii** organelles are only found in the cells of multicellular organisms.

2.2 Getting to know your microscope

Science understanding

FOUNDATION | STANDARD | ADVANCED

1 Name the parts labelled A to G on the diagram using terms from the box below.

coarse focus knob	eyepiece	fine focus knob	handle	light source	objective lens	stage

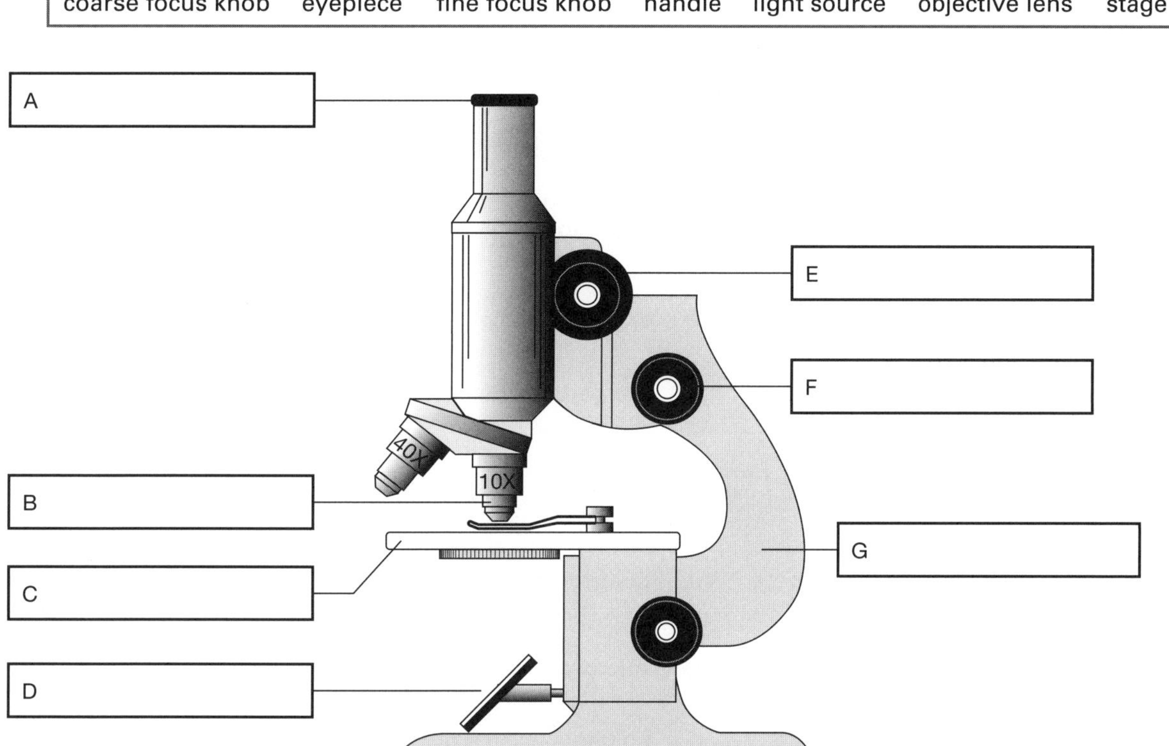

2 Identify the description of each of the following parts of the microscope to the correct term by joining them with a line.

(a) eyepiece	part of the microscope on which the specimen is placed
(b) coarse focus knob	sharpens the focus on high power
(c) stage	unit used to measure microscopic objects
(d) objective lens	equipment used to make a wet mount
(e) micrometre (μm)	the object being studied using the microscope
(f) specimen	the lens of the microscope closest to the specimen
(g) light source	the part of the microscope you look through
(h) fine focus knob	used to light the specimen
(i) slide and coverslip	used to focus the microscope on low power

RATE MY UNDERSTANDING
Shade the face that shows your rating 

2.3 Plant and animal cells

Science understanding

FOUNDATION | **STANDARD** | ADVANCED

1 Plant and animal cells have a number of different features. Match the cell structure (on the left-hand side) with the job it does (on the right-hand side) by joining them with a line. One has been done for you.

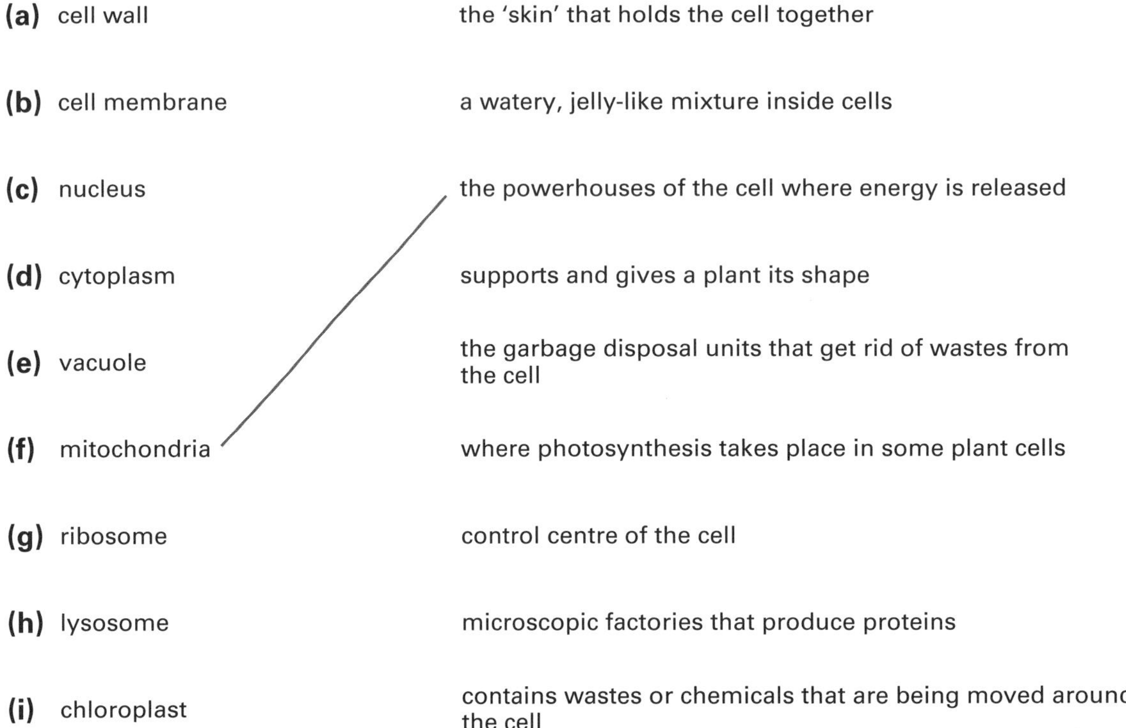

2 Some features are found only in plant cells, and others are common to both plant and animal cells. Identify the structures of the plant and animal cells by selecting the correct term from the box.

cell membrane	cell wall	chloroplast	cytoplasm	nucleus	vacuole

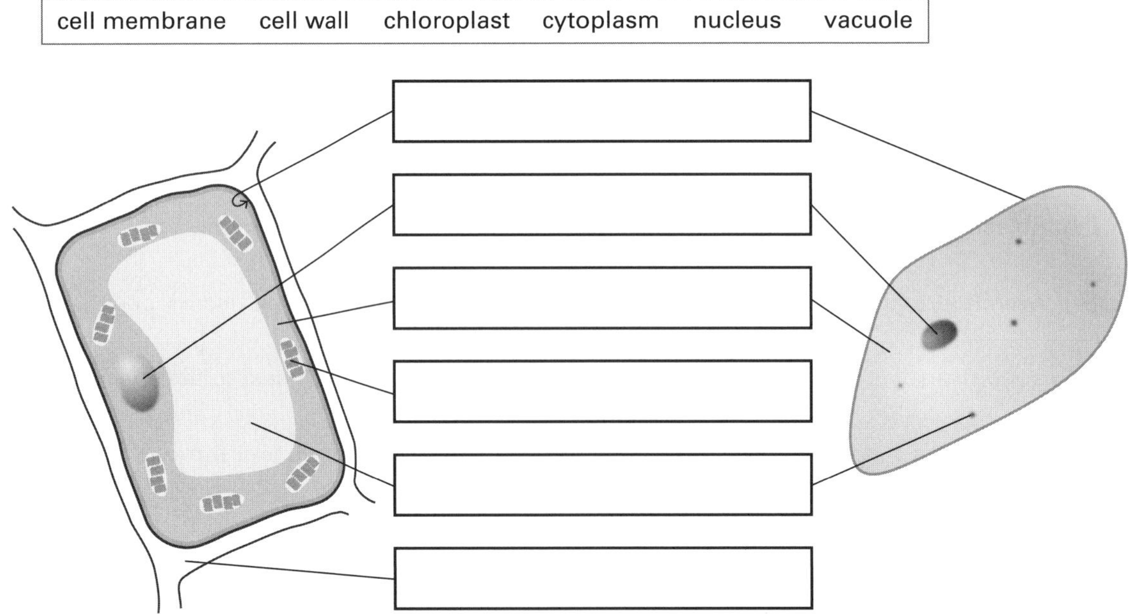

RATE MY UNDERSTANDING
Shade the face that shows your rating

2.4 Microscopic field of view

Science inquiry skills

FOUNDATION	STANDARD	ADVANCED

When you look at an organism or part of an organism through a microscope, you can estimate the size by knowing the diameter of the microscope's field of view. The field of view is the circle you see when you look through a microscope. The diameter of the field of view can be calculated by using a mini-grid. A mini-grid is a plastic grid divided into millimetre squares. It is like clear plastic graph paper.

diameter (*n*) a straight line passing from one side of a figure to the other side through the centre

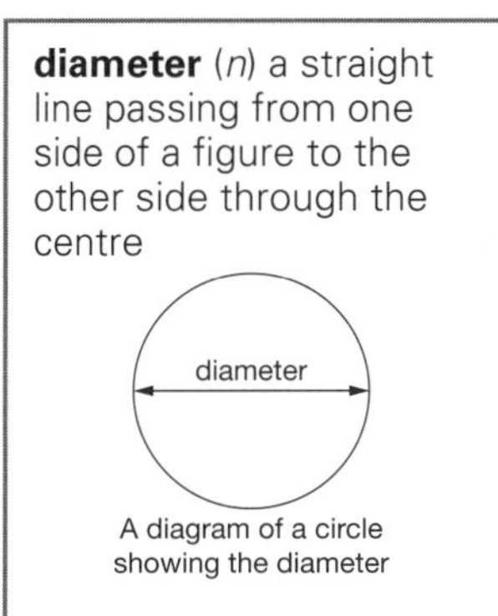

A diagram of a circle showing the diameter

1 The following diagrams show the field of view for three different magnifications of the same microscope. Work out the diameter of the field of view for each of the magnifications to the nearest whole millimetre (mm) and micrometre (μm). Remember that 1 mm = 1000 μm

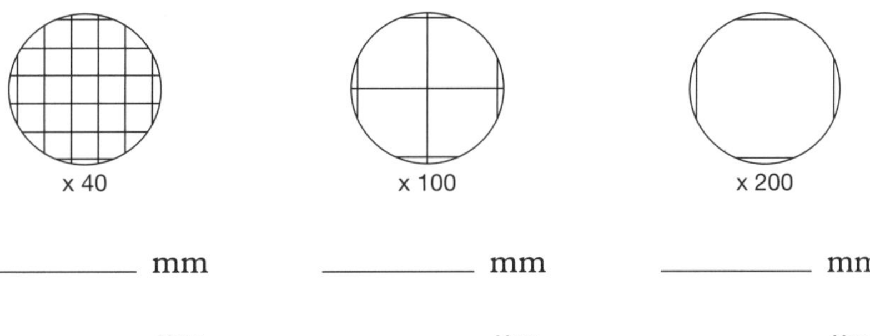

__________ mm   __________ mm   __________ mm

__________ μm   __________ μm   __________ μm

Use the diagrams and information, to answer the following questions.

2 Explain what happens to the diameter of the field of view as the magnification increases?

____

(3) Estimate how large the cell in the diagram below is in millimetres. This cell was observed at a magnification of × 40 on the same microscope as for question 1.

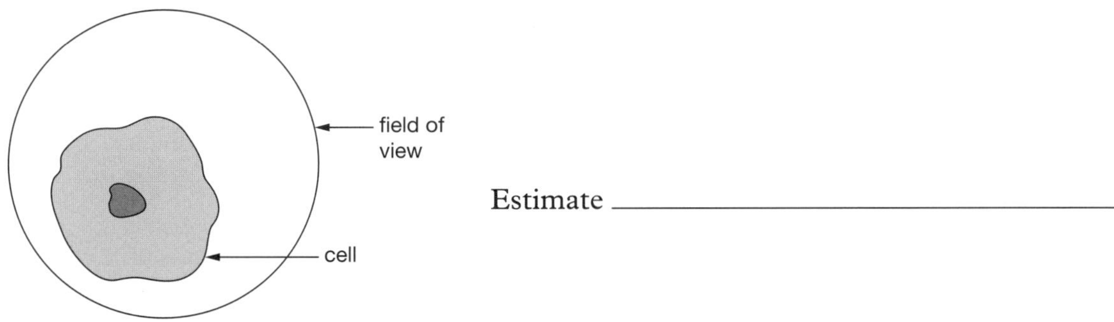

Estimate ____________________________________

(4) Paramecium are small unicellular freshwater organisms that look like this:

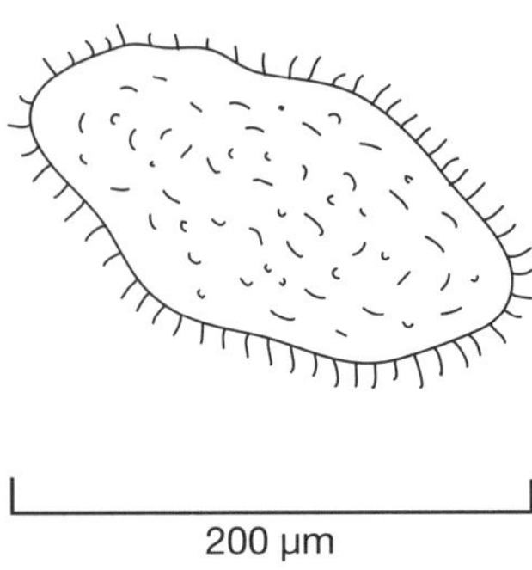

They are about 200 µm long and swim through the water using fine hairs (cilia) on the outside of their bodies. Draw a picture of a paramecium as it would appear at a magnification × 200 in the circle below using the same microscope as in Question 1.

5 Paramecium swim so rapidly they can be difficult to observe because they move out of the field of view. Propose a way to slow them down so they are easier to examine.

____

____

____

RATE MY UNDERSTANDING
Shade the face that shows your rating  

2.5 Size of cells

Science inquiry skills

FOUNDATION | **STANDARD** | ADVANCED

Processing & Analysing | Communicating

Cells are generally microscopic and many are only fractions of a millimetre. It is not possible to measure the size of cells using millimetres. The unit used is the micrometre. A micrometre is a thousandth of a millimetre and has the symbol μm.

1. Calculate the missing values in the table below by converting to the units shown. To convert from centimetres to millimetres, multiply by 10. To convert millimetres to micrometres, multiply by 1000. To reverse each of these, divide by these factors of 10 and 1000. The first one has been done for you.

Centimentres to millimetres to micrometres conversion table

cm ×10 → mm ×1000 → μm; μm ÷1000 → mm ÷10 → cm

cm	mm	μm
0.03	0.03 × 10 = 0.3	0.03 × 1000 = 300
0.7		
	2	
		45 × 1000 = 45
0.03		
		130
	0.04	
		78

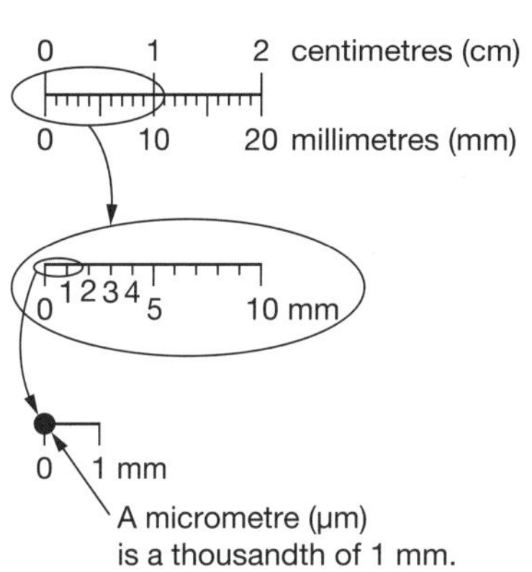

Using scales

2. When scientists draw diagrams of very small objects, they enlarge them. A scale is then added to the diagram to give an idea of the real size. Below is an enlarged drawing of a dust mite and its scale. Dust mites are found everywhere but they are too small for us to see easily. Follow the steps below to work out how big the dust mite really is.

Step 1 Take a piece of paper and place it along the scale next to the drawing

Step 2 Mark off sections along the paper that are exactly the same distance apart as the scale. Number the sections to make a ruler.

Step 3 Use your paper ruler to measure the length of the body from point A to point B. This measurement is 30 micrometres (μm). The actual length of the dust mite is 30 micrometres.

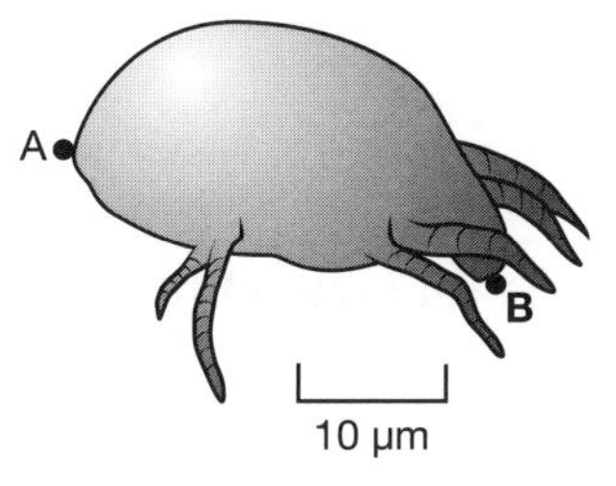

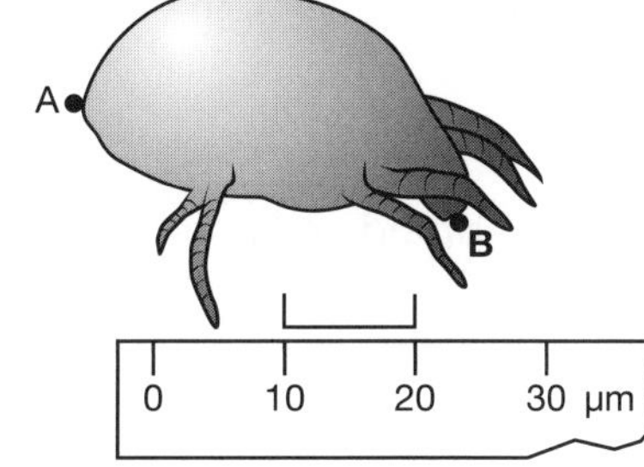

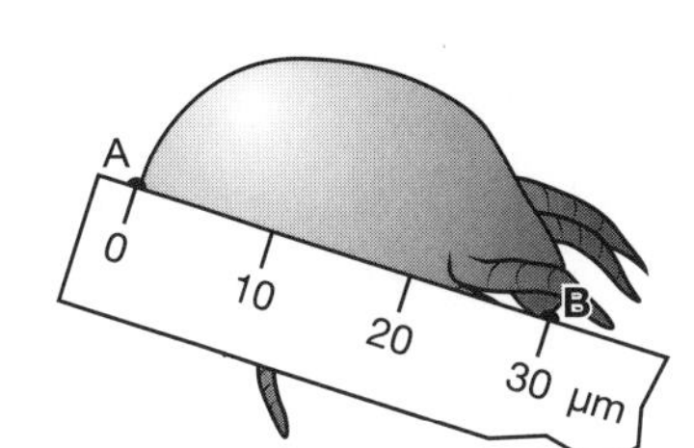

2.5 Size of cells

(a) Use these enlarged drawings to calculate how big the cells really are then complete the table below.

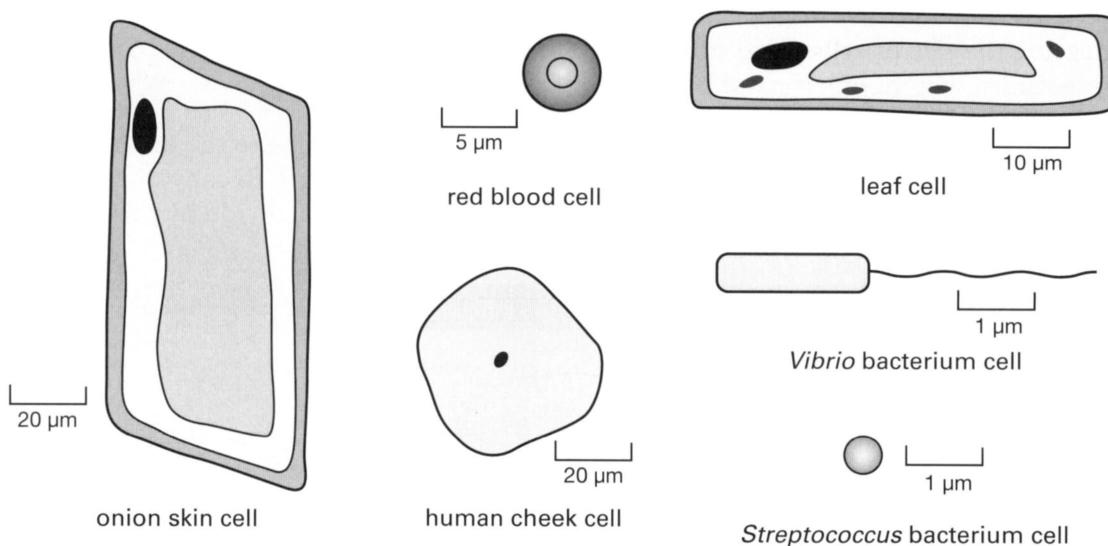

Question	Actual size (µm)
What is the diameter of the red blood cell?	
What is the diameter of the human cheek cell?	
What is the length of the leaf cell?	
What is the width of the leaf cell?	
What is the length of the onion skin cell?	
What is the width of the onion skin cell?	
What is the length of the body of the *Vibrio* bacterium cell?	
What is the width of the body of the *Vibrio* bacterium cell?	
What is the length of the tail of the *Vibrio* bacterium cell?	
What is the diameter of the *Streptococcus* bacterium cell?	

(b) Calculate the number of *Streptococcus* bacterial cells placed side by side that would fit across the diameter of a human cheek cell.

____

(c) Calculate the number of red blood cells placed side by side that would fit along the length of a leaf cell.

____

RATE MY UNDERSTANDING
Shade the face that shows your rating

2.6 Electron microscope images

Science inquiry skills

FOUNDATION	STANDARD	ADVANCED

Processing & Analysing | Evaluating

When scientists first used electron microscopes many familiar structures looked very different to how they appeared under a light microscope. They also saw things that had never been seen before. They had to try to make sense of the images.

The following pictures are electron micrographs of familiar objects.

1 **(a)** Match photographs **A** to **D** with the correct name from the box. Write your answers on the labels underneath the images.

eye of a fly	salt crystals	sun flower pollen	paper

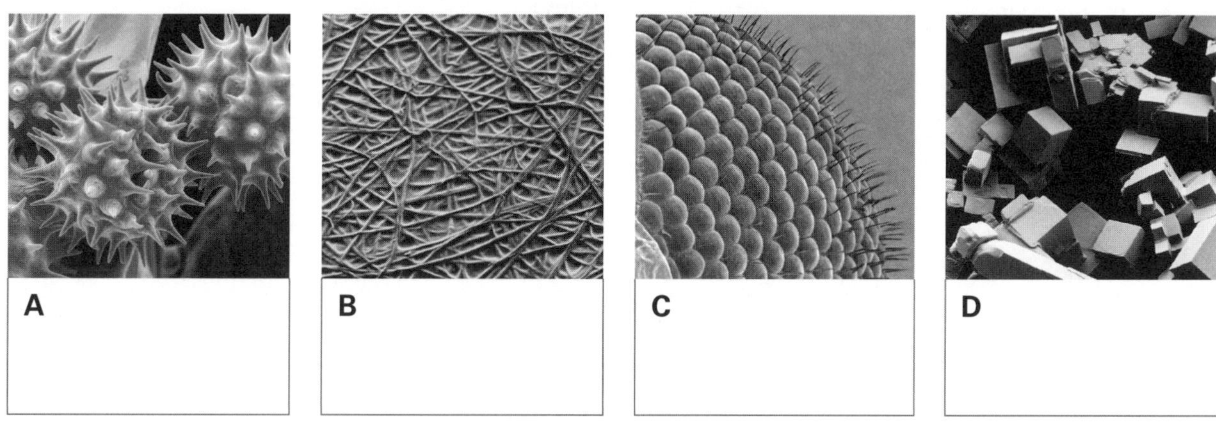

(b) Suggest what photographs **E** to **H** represent. Write your ideas under each image.

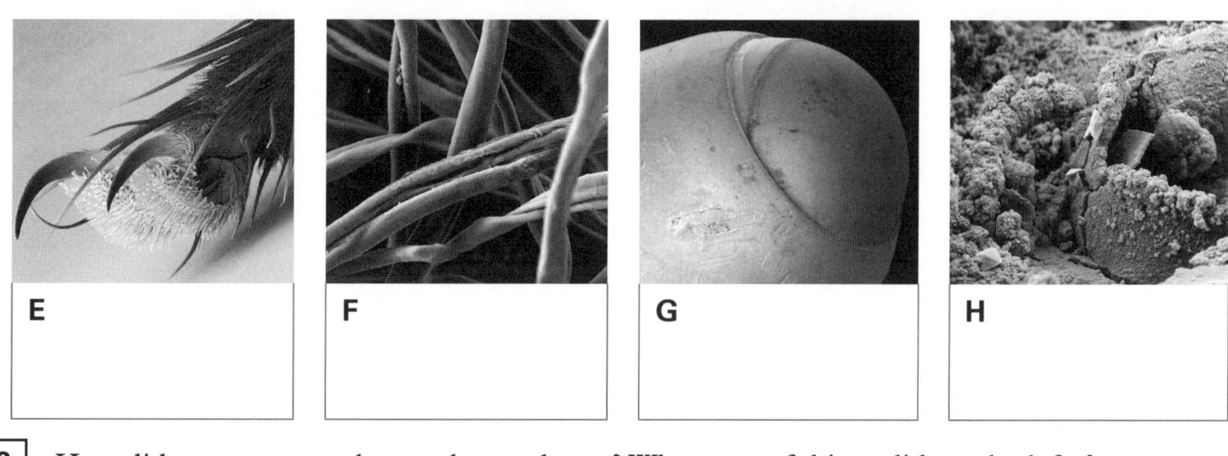

2 How did you try to work out what each was? What type of things did you look for?

RATE MY UNDERSTANDING
Shade the face that shows your rating

2.7 The shape and structure of cells

Science inquiry skills

FOUNDATION	**STANDARD**	ADVANCED

The cells found in plants and animals are of different shapes and sizes depending on what they do.

1. Think about where these cells are found and the jobs they have to do. Explain why their shape means that the cells are well suited to their jobs.

Cell type and function suitability

Cell type	Diagram of cell	Function	Why the shape makes the cell suited to its job
(a) human skin cells		provides a complete covering for the body	
(b) nerve cells in brain		sends information to and receives information from different parts of the brain	
(c) nerve cell in body		sends information from all parts of the body to the brain	
(d) cell from small intestine		passes digested food from space inside the intestine into the circulatory system of the body	

digested (*v*) food broken down in the stomach

2. Some plant cells change as they get older. These three diagrams represent a cell from the stem of a tree.

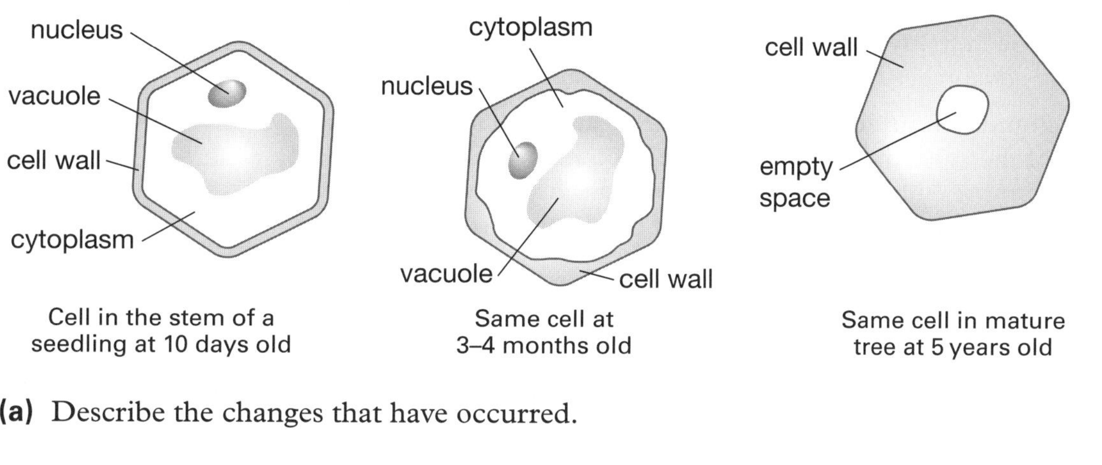

Cell in the stem of a seedling at 10 days old

Same cell at 3–4 months old

Same cell in mature tree at 5 years old

(a) Describe the changes that have occurred.

____

____

(b) How do these changes help the plant survive?

____

____

3. The following table contains a list of features of particular cells. Suggest how each feature helps the cell carry out its job.

Cell features and functions

Feature of cell	How the feature helps the cell do its job
(a) Cells in the upper layers of leaves have large numbers of chloroplasts.	
(b) Muscle cells in the human leg have large numbers of mitochondria.	
(c) Cells in plant stems that carry water from the roots are joined end to end so they form a continuous tube.	
(d) Bone cells make a structure which can store minerals. This makes a hard substance that surrounds the bone cells.	

RATE MY UNDERSTANDING
Shade the face that shows your rating

2.8 Surface area

Science inquiry skills

FOUNDATION | STANDARD | **ADVANCED**

Processing & Analysing | Questioning & Predicting | Evaluating

exchange (*n*) a swap or transfer of something
membrane (*n*) a thin layer that acts like a skin

Cells come in many different shapes. This activity explores the possible advantages of cells of different sizes and shapes.

The cell membrane acts as a barrier between the outside and the inside of the cell. Anything that the cell requires or needs to get rid of has to move out through the membrane. The cell will function best if there is an efficient exchange of materials across the membrane.

1 Imagine a cell as being like a cube.

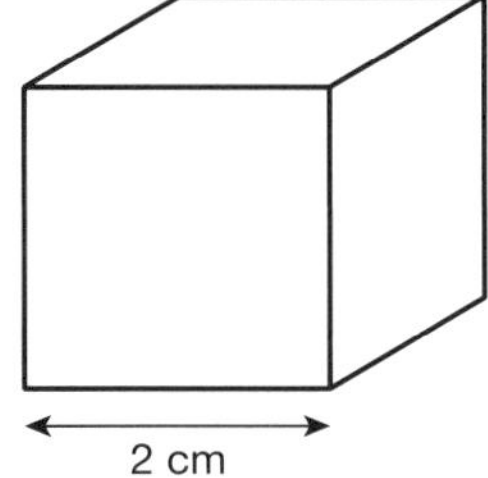

The surface of this cube is made up of the faces of the cube. The surface area of the cube is the area of all six sides of the cube added together. This cube has sides that are each 2 cm long.

To calculate the surface area of the cube:

Work out the surface area of one side of cube.

Area = Length × Width

Area = 2 × 2

Area = 4 cm^2

Multiply the surface area of one side (4 cm^2) by the number of sides the cube has (6)

Total surface area = 4 × 6

Total surface are = 24 cm^2

(a) Imagine the cube now being cut into eight smaller cubes.

Calculate the surface area of one of the smaller cubes and then the total surface area of all eight cubes combined.

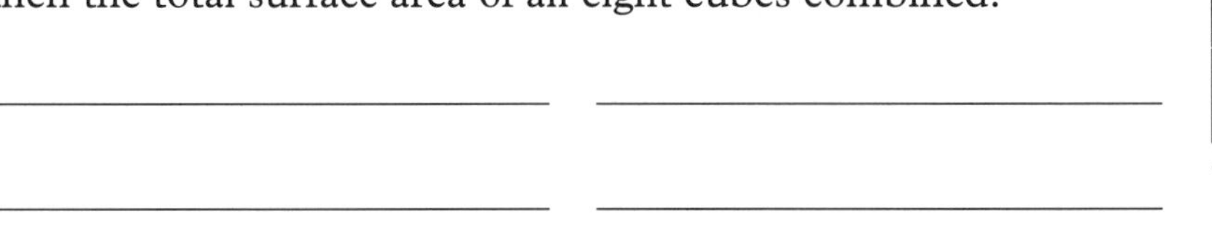

1 cm

(b) Now cut each small cube into eight smaller cubes again.

Calculate the surface area of one of the smaller cubes and then the total surface area of all the cubes.

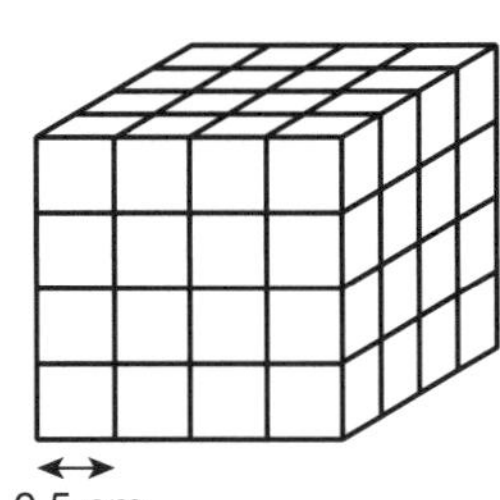

(c) Record your results for (a) and (b) in rows 2 and 3 of this table.

Length of side (cm)	Surface area of 1 cube (cm^2)	Number of cubes	Total surface area (cm^2)
2	4	1	24
1		8	
0.5			
0.25			
0.125			

(d) Describe the pattern of change in the surface area by looking at the results in the table.

(e) Use this pattern to predict the values that will complete the last two rows of the table. At each stage, each cube is cut into eight. Write your predictions in the table.

(f) Calculate how much faster water would move into 64 cube-shaped cells with sides of 0.5 cm than into one cell with sides of 2 cm. (Hint: compare the total surface area of each.)

2. The cells covering the surface of a plant root are mostly like those shown in diagram A to the right. However, in the area where most water enters the root, the cells are shaped like those of diagram B. Explain why these root cells would be an advantage to the plant.

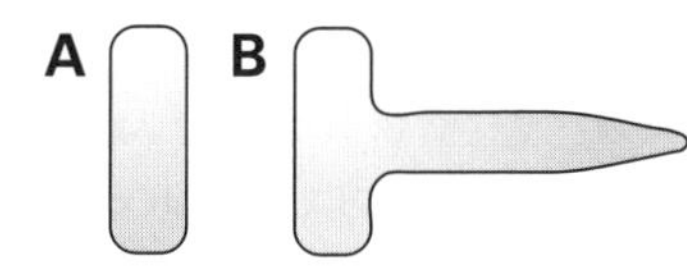

3. Diagram A to the right shows the typical shape of the cells lining your gut. In a part of the gut called the small intestine the cells are more like those of diagram B. Suggest what the purpose of the small intestine might be.

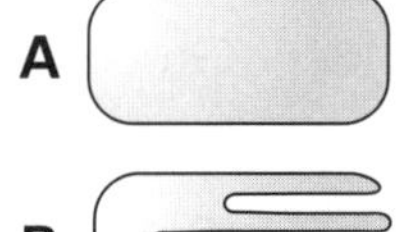

4. During the night, leaf cells take in oxygen gas from the atmosphere. They also release carbon dioxide gas. Compare the two arrangements of cells shown in diagrams A and B to the right and decide which one would exchange gas more efficiently. Explain your answer.

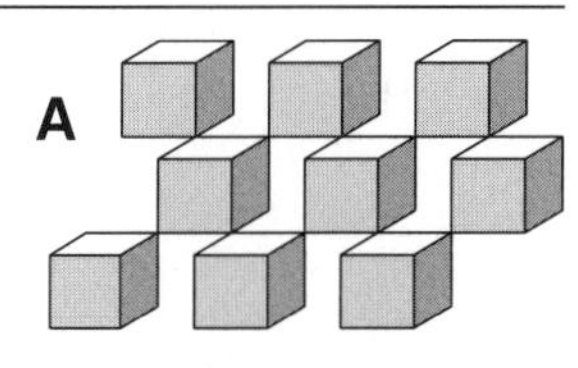

RATE MY UNDERSTANDING
Shade the face that shows your rating

2.9 Functioning plant

Science inquiry skills

FOUNDATION	STANDARD	ADVANCED

Processing & Analysing **Evaluating**

chloroplasts (*n*) organelles in green plants that contain chlorophyll and where photosynthesis takes place

distributed (*v*) spread out

efficiency (*n*) ability to do a job effectively

raw materials (*n*) basic elements/ingredients, not processed

waste product (*n*) unwanted material or substance

1 Use words from the box below to complete the sentences.

carbon dioxide	chlorophyll	glucose	oxygen
phloem	root hairs	sunlight	water
xylem			

(a) The raw materials for photosynthesis are ______________ and ______________ .

(b) Photosynthesis also needs the green chemical called ______________ found in chloroplasts together with energy from ______________ .

(c) ______________ is produced by photosynthesis.

(d) A waste product from photosynthesis is ______________ .

(e) Water is taken into the plant through ______________ and travels up the ______________ to the leaves of the plant.

(f) Glucose is distributed throughout the plant in the ______________ .

(2) Why is it an advantage for plants to have cells with large numbers of chloroplasts near the upper surface of the leaf?

____

____

____

3 Glucose is carried in the phloem from the cells where it is made, up the plant towards the tips of branches and down towards the roots. Explain why glucose is needed in:

(a) the tips of branches

____

____

(b) in the roots.

____

____

4 **(a)** Contrast the organisation of the tissues in the two leaves shown in diagrams A and B. Write these differences in the first column of the table below in question 4(b).

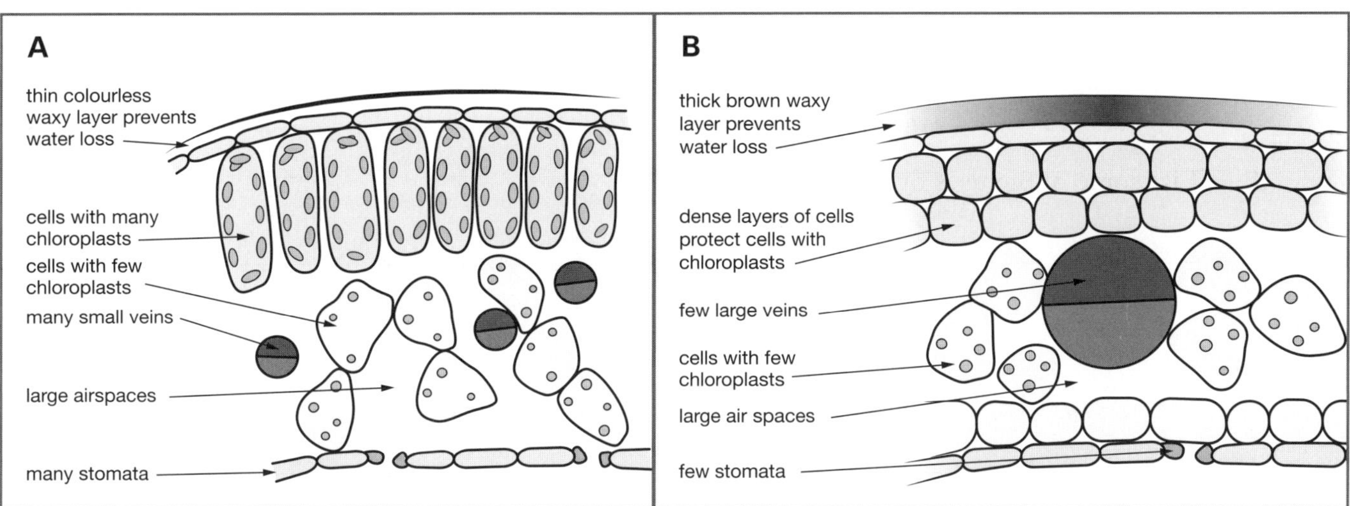

(b) What effect would the differences between the two leaves, A and B, have on the leaves' ability to carry out photosynthesis efficiently? Write your answers in the second column of the table.

Leaf organisation differences	
Difference	**Effect on photosynthesis**

2.10 Growing cells

Science as a human endeavour

FOUNDATION | **STANDARD** | ADVANCED

1 Define the term *cell culture*.

____

2 List some of the uses of cultured cells.

____

____

(3) Explain how stem cells are different from other cells.

____

____

(4) Below is a jumbled set of steps for growing a bladder for transplant to a human. Read each step and arrange the steps in the order the process occurs by numbering the steps 1 to 5.

- ☐ Make a scaffold that cell culture can grow on in the shape of a bladder.
- ☐ Grow the cells taken from the patient in a cultured solution.
- ☐ Transplant the cell culture back into the body of the patient where they continue to grow.
- ☐ Take cells from the bladder lining and from muscle of the patient.
- ☐ When enough cultured cells have grown, place them on a shell-shaped scaffold to grow more cells in a cultured solution.

A bladder grown using cultured cells.

(5) Why is a shell-shaped scaffold necessary when growing a new organ like a bladder but is not used when growing skin?

____

____

(6) What are two types of cells that were cultured to grow the new bladder?

____

____

RATE MY UNDERSTANDING
Shade the face that shows your rating

2.11 Literacy review

Science understanding

FOUNDATION	STANDARD	ADVANCED

1 (a) Use the clues to complete the crossword. All words are written across the crossword.

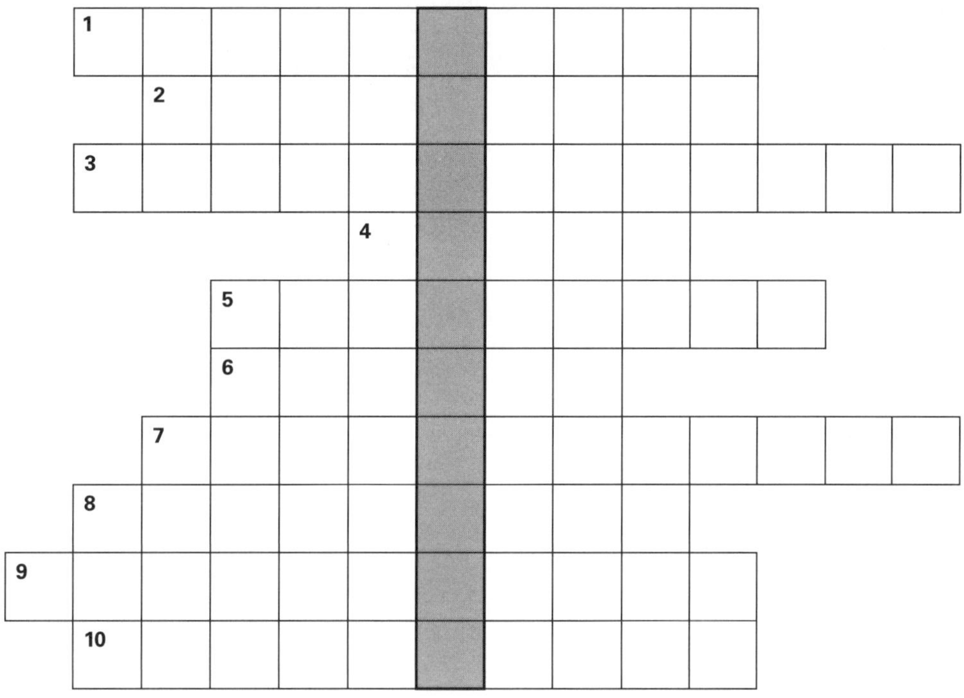

1. Unit used to measure microscopic things
2. When small things are made to look bigger than they are
3. Describes an organism made up of many cells
4. Group of different tissues that work together
5. Watery, jelly-like substance found inside cells
6. Groups of cells of the same type
7. Powerhouses of the cell
8. Organelles that produce proteins
9. The organelle that makes plants green and where they make their food
10. Small parts found within cells

(b) Read the letters in the shaded column. State the word they spell.

____

(c) Define this word.

____

____

RATE MY UNDERSTANDING
Shade the face that shows your rating  

2.12 Thinking about my learning

Consider the statements in the table below and rate your confidence in your knowledge, understanding or skills for each statement by placing a tick (✔) in one of the columns.

Statements	I am not sure.	I understood some things.	I understood most things.	I understand this really well.
I can label the parts of a microscope				
I can operate a microscope correctly				
I can prepare a wet mount to observe a specimen				
I can calculate the magnification on a microscope				
I can correctly identify the difference between a plant and animal cell				
I can identify the different organelles in a cell				
I understand the function of a cell membrane				
I understand the function of the nucleus				
I can convert millimetres to micrometres				
I understand the function of a chloroplast				
I am confident I know the difference between a unicellular and multicellular organism				
I am confident I can name a variety of specialised cells				
I can name a variety of different tissues in animals				
I can identify different organs in the human body				
I understand the purpose of mitosis				
I can analyse data in a table or graph				

CHAPTER 3

Body systems

3.1 Knowledge preview

Science understanding

FOUNDATION	STANDARD	ADVANCED

1 The human body is made up of many bones. Use the words listed in the box below to correctly label the human skeleton in Figure 3.1.1.

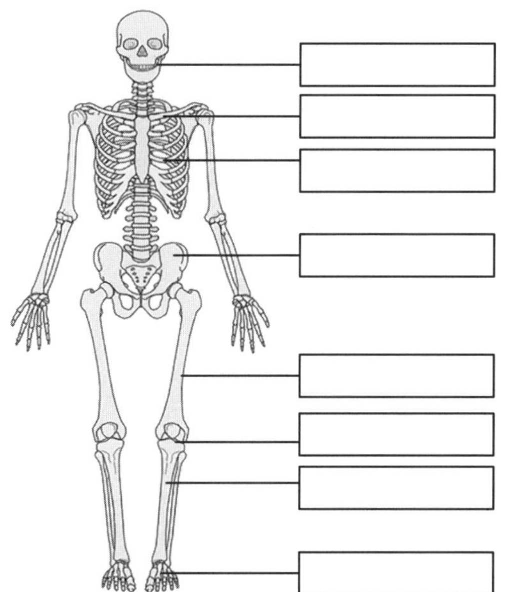

Figure 3.1.1 Human skeleton bones

2 Each of the descriptions of structures on the left table matches a name on the right. Write the correct number in the middle column that matches the name of the structure to its description.

Description of structures	Matching number	Names of structures
A connects bone to bone		**1** tendon
B helps the chest to expand		**2** lung
C calcium is important for this		**3** heart
D absorbs nutrients		**4** vein
E filters wastes from the body		**5** diaphragm
F carries blood back the heart		**6** ligament
G connects muscle to bone		**7** stomach
H is composed of 4 chambers		**8** kidney
I is made of tiny sacs		**9** skeleton
J produces acid for digestion		**10** small intestine

3 Figure 3.1.2 shows the major organs in the human body.

(a) Label the major organs on the diagram.

(b) Under the name of each organ, write one fact that you know about that organ. The fact may relate to the function or structure of the organ, any related diseases, medical procedures or technological advances related to the organ.

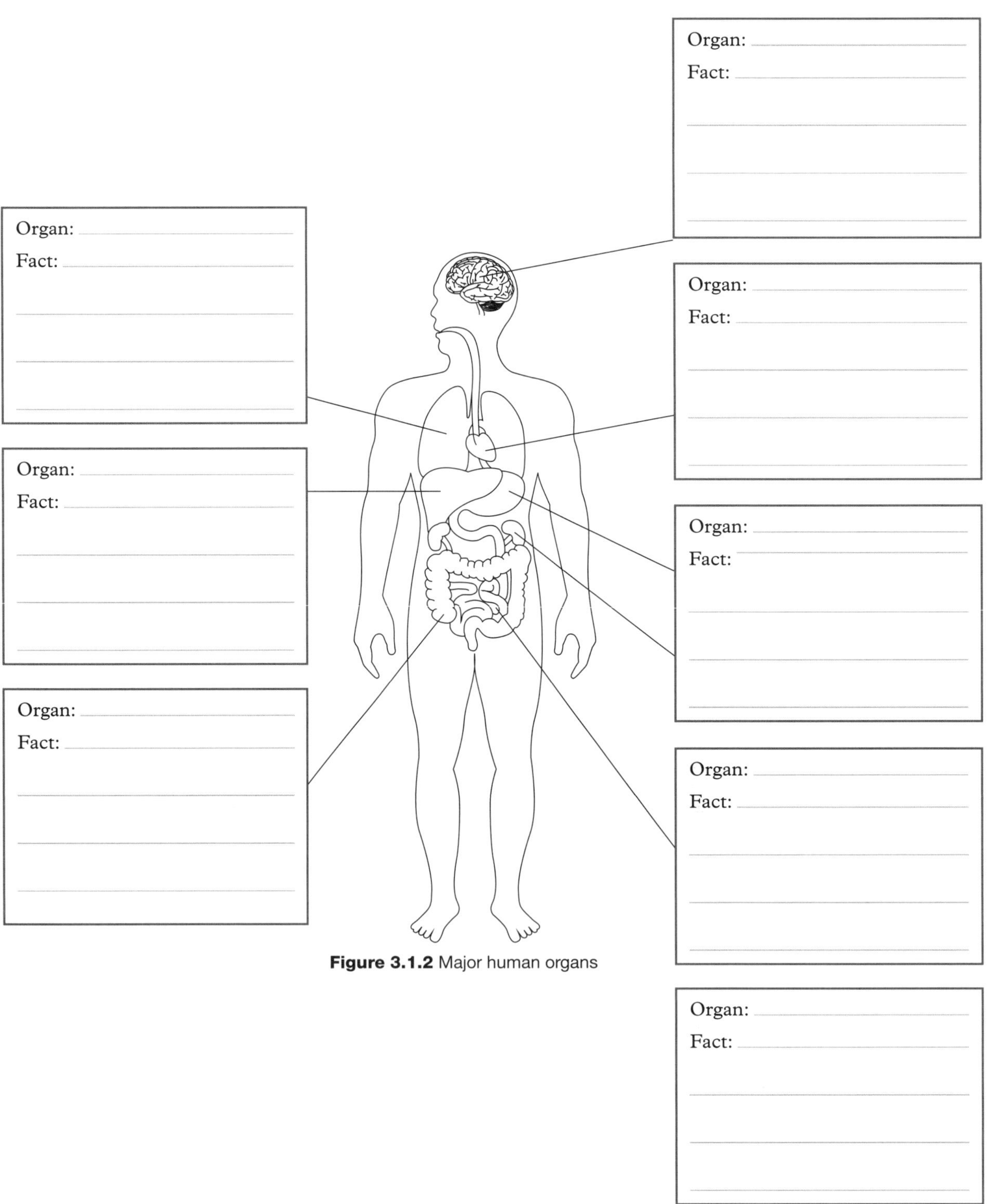

Figure 3.1.2 Major human organs

Organ:

Fact:

3.2 Parts of the digestive system

Science understanding

FOUNDATION	STANDARD	ADVANCED

Your digestive system changes the food you eat into a form your body can use.

1 Figure 3.2.1 shows the human digestive system. Identify the following parts by colouring each as follows and adding labels:

small intestine: red

oesophagus: blue

large intestine: green

stomach: yellow

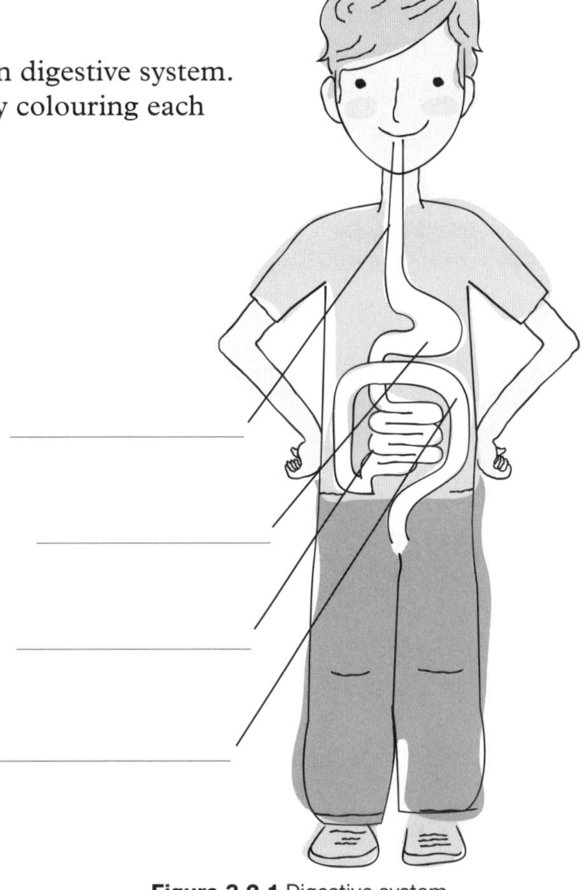

Figure 3.2.1 Digestive system

2 Match the part of the digestive system (on the right) with its description (on the left) by drawing lines between them.

Part of the digestive system
mouth
oesophagus
stomach
small intestine
large intestine

Description
This is where water is taken back into the body and any wastes and unwanted food are passed out of the body through the anus. This structure is short but wide.
Most of the digestion is finished here. Food is now very tiny particles that can be absorbed by the body. This structure is long, but narrow. Useful nutrients pass through the wall into the body where they are taken by the blood to the cells.
Mechanical digestion starts here when you chew your food. Chemical digestion of carbohydrates starts here using chemicals found in saliva.
This is the tube that carries the chewed food from the mouth to the stomach. A muscle wave known as peristalsis moves the food to the stomach.
This is where very strong acid helps to digest proteins and helps to kill any bacteria in the food.

RATE MY UNDERSTANDING
Shade the face that shows your rating

3.3 Mechanical and chemical digestion

Science inquiry skills

FOUNDATION	STANDARD	ADVANCED

Processing & Analysing

When you bite off a piece of apple and chew it into smaller pieces, this is mechanical digestion. Chemical digestion occurs when the complex sugars in the apple are changed into simple sugars by chemicals in your mouth and small intestine.

The following five diagrams represent digestion. Name the type of digestion each represents. Justify your decision in each case.

1 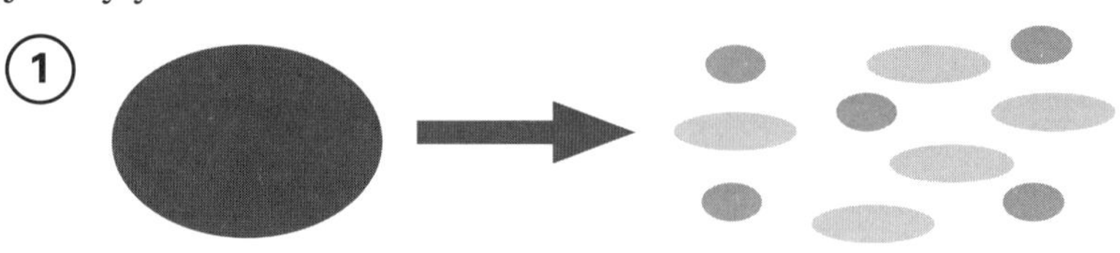

____

____

2 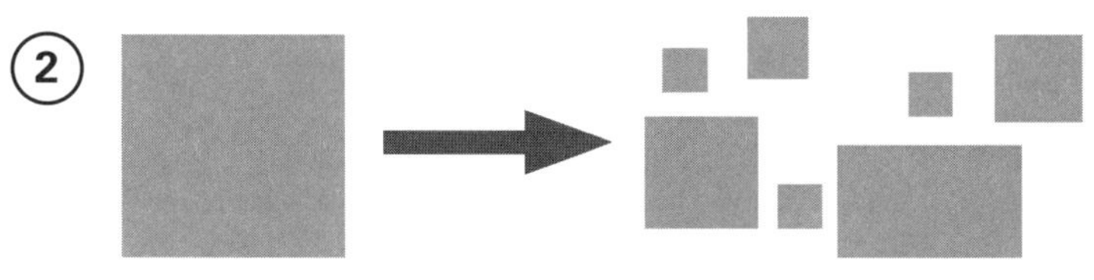

____

____

3 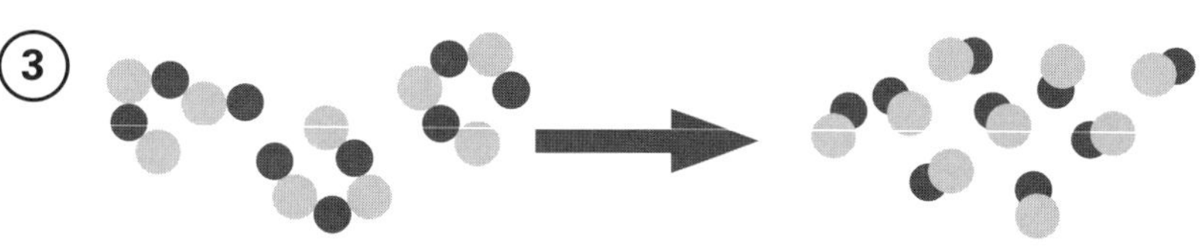

____

____

4 

____

____

5 

____

____

RATE MY UNDERSTANDING
Shade the face that shows your rating   

3.4 Investigating villi

Science inquiry skills

FOUNDATION	STANDARD	ADVANCED

Processing & Analysing

The following diagrams represent three different surfaces X, Y and Z.

1 Use cotton thread or fine string and a ruler to measure the length of the lower side of each surface. The Hint box below shows you how to do this.

Length of surface X ____________ Y ____________ Z ____________

Hint

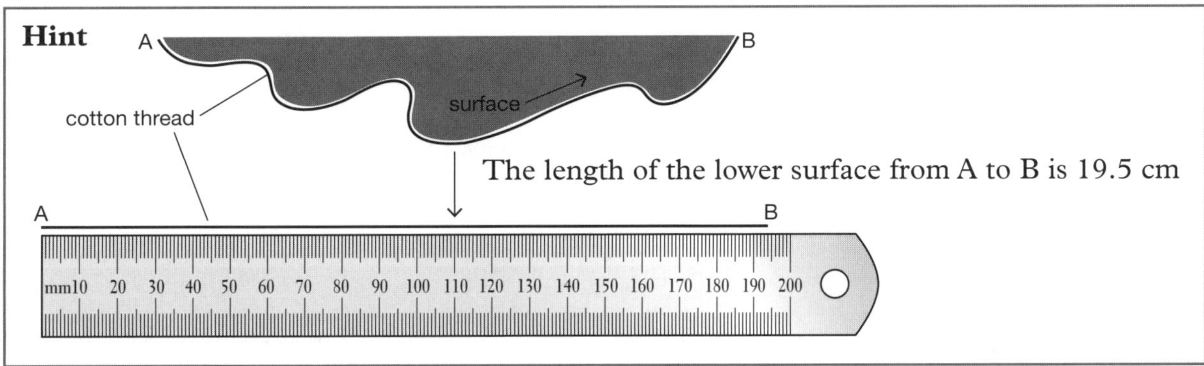

(2) If each centimetre of surface absorbs 5 mL of digested material every 10 minutes, how much digested material is absorbed in one hour by each of the surfaces X, Y and Z. A worked example is provided to help you with your calculations.

X ____________________

Y ____________________

Z ____________________

Worked example:

- If the surface length is 10 cm
- And 5 mL is digested per cm
- Then 50 mL (10 cm × 5 mL) is digested every 10 minutes

As there are 60 (10 min × 6) minutes in an hour

- Then over an hour the total amount digested is 50 mL × 6 = 300 mL

(3) Compare the effectiveness of surfaces X, Y and Z in digesting food.

____

____

(4) Explain why it is an advantage to have villi lining the small intestine.

____

____

RATE MY UNDERSTANDING
Shade the face that shows your rating

3.5 The respiratory system

Science understanding

FOUNDATION	STANDARD	ADVANCED

1 Label the parts of the respiratory system indicated in Figure 3.5.1 by selecting the correct term in the box below.

alveoli
bronchiole
bronchus
diaphragm
lung
trachea

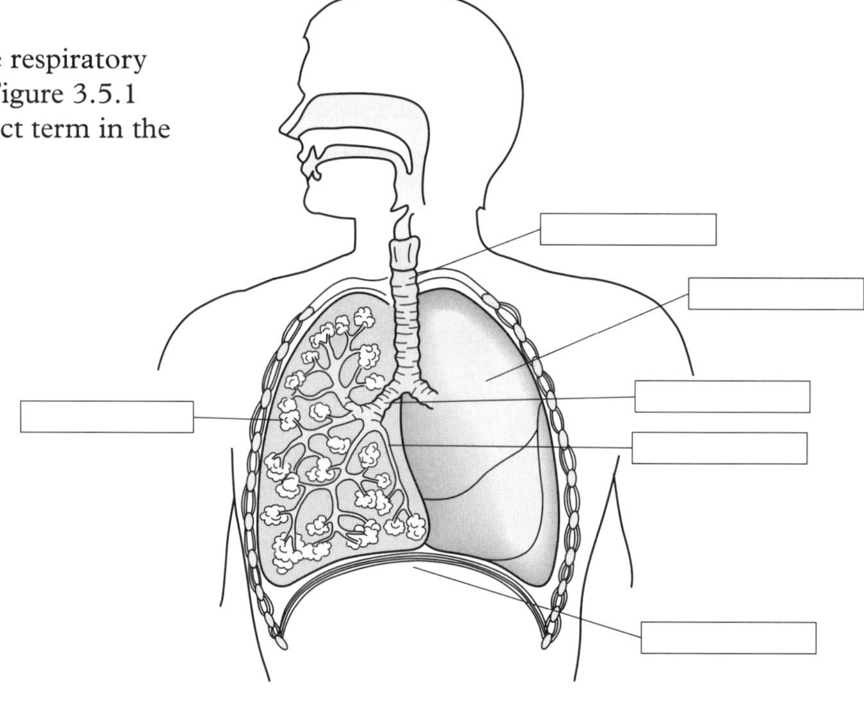

Figure 3.5.1 Respiratory system

2 Draw lines from the parts of the respiratory system in the left-hand column to match with their descriptions in the right-hand column.

Parts of the respiratory system
trachea
bronchi
alveoli
diaphragm

Description
A sheet of muscle that separates the chest from the abdomen. It contacts and flattens as you breathe in and domes up as you breathe out.
A cluster of sacs in which oxygen and carbon dioxide are exchanged.
Thin-walled tube reinforced with rings of cartilage. You can feel these rings as ridges on the front of your throat.
One of these carries air into each lung.

(3) Air is breathed in by the respiratory system. Within the lungs, exchange of gases takes place and the air that is breathed out has a different composition.

Gas	Percentage (%)	
	Inhaled air	Exhaled air
nitrogen	78	78
oxygen	21	17
inert gases such as argon	1	1
carbon dioxide	0.04	4
water vapour	little	saturated

(a) List the gases that are present in the same quantities in inhaled and exhaled air.

____

(b) Explain why the quantities of these gases do not change.

____

____

____

RATE MY UNDERSTANDING
Shade the face that shows your rating   

3.6 The heart

Science inquiry skills

FOUNDATION	STANDARD	ADVANCED

Processing & Analysing | Communicating

1 A diagram of the heart (Figure 3.6.1) is provided below.

(a) Add labels from the box to identify the parts of the heart.

(b) Lightly colour the heart and blood vessels to identify where there is oxygenated blood (red) and deoxygenated blood (blue).

(c) Add arrows to identify the direction of blood flow through the heart.

(d) At the end of the blood vessels, name the part of the body the blood is flowing to or coming from.

aorta
left atrium
left ventricle
pulmonary artery
pulmonary vein
right atrium
right ventricle
valves
vena cava

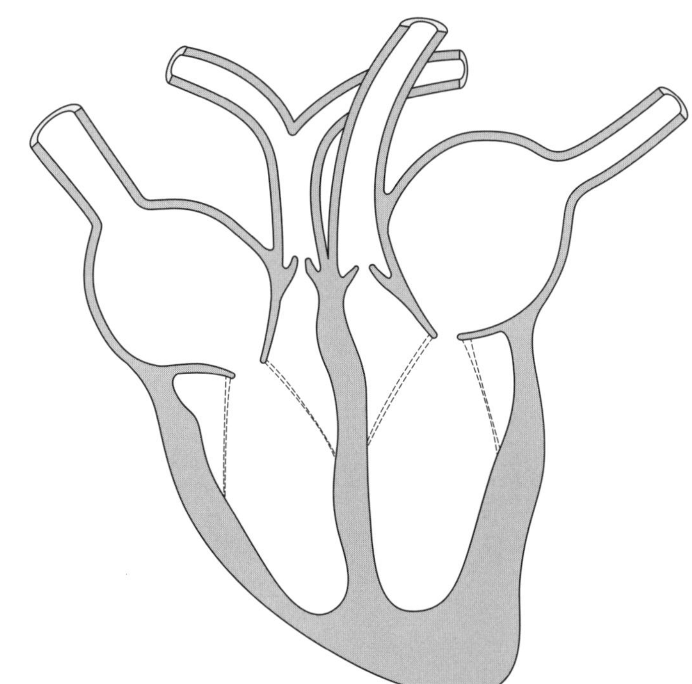

Figure 3.6.1 The human heart

2 **(a)** Identify whether the right or left ventricle is larger and has thicker walls.

____

(b) Explain why this ventricle has thicker walls.

____

____

(3) Construct a flow diagram to show the passage of the blood through the body and heart. Start and end with the right ventricle.

RATE MY UNDERSTANDING
Shade the face that shows your rating

3.7 Effect of exercise on pulse rate

Science inquiry skills

FOUNDATION	**STANDARD**	ADVANCED

Communicating

Mary and Anika had the change in their heart rate recorded during exercise. Mary trained on a regular basis and was reasonably fit. Anika did not train at all. The results are shown below.

axes (*n*) plural of axis. The vertical and horizontal lines that show the units on a graph.

Pulse rate during exercise

When recorded	Time (minutes)	Pulse rate (beats per minute)	
		Mary (fit)	Anika (unfit)
before exercise	1	55	62
	2	56	61
	3	55	62
	4	55	62
	5	56	61
during exercise	6	60	70
	7	70	80
	8	75	90
	9	97	120
	10	106	130
	11	120	140
	12	130	142
	13	132	148
	14	131	150
	15	131	150
after exercise	16	115	145
	17	98	140
	18	75	130
	19	60	115
	20	55	100

3.7 Effect of exercise on pulse rate

1. Use the data to plot the graph on the axes provided. Use one colour for Anika and a different colour for Mary. Include a key with your graph.

Mary's and Anika's pulse rate comparison

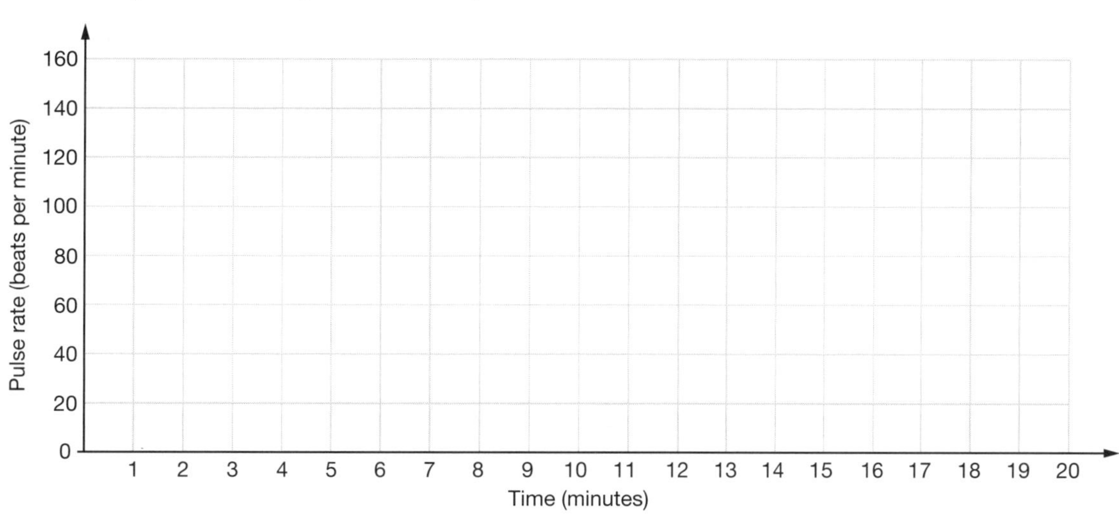

2 Describe the pulse rates for Mary and Anika during exercise.

____

____

(3) Explain why these changes occurred.

____

____

____

(4) Compare the changes in pulse rate and account for the differences at each of the following times:

(a) before exercise

____

(b) in the first 5 minutes of exercise

____

(c) in the second 5 minutes of exercise

____

(d) after exercise was completed.

____

RATE MY UNDERSTANDING Shade the face that shows your rating			

3.8 The excretory system

Science inquiry skills

FOUNDATION	STANDARD	ADVANCED

Processing & Analysing | Evaluating

Your body produces solid, liquid and gaseous waste materials. Wastes must be removed from the body or you will become ill. The digestive system produces waste in the form of faeces. This solid waste forms from undigested food. The excretory system does not include solid wastes. The excretory system refers to elimination of liquid and gaseous wastes from the skin, lungs and kidneys.

1. Complete the table by filling in:
   - the main function of each organ that excretes wastes
   - the types of waste produced
   - what happens when the system is not working normally.

   You may do some research in order to complete this task.

Major organs of excretory system		
Lungs	function of organ	
	type/s of waste	
	system not working well	
Skin	function of organ	
	type/s of waste	
	system not working well	
Kidneys	function of organ	
	type/s of waste	
	system not working well	

2 Albumin is a protein that nourishes tissues, stops fluid leaking out of blood vessels and transports substances like calcium and vitamins around the body. Healthy kidneys will prevent albumin passing into urine and will retain it in the blood. The more damaged the kidney, the more albumin that will be excreted in urine.

(a) The diagrams below model two different kidneys. In the space provided below each diagram:

- Identify the healthy and the damaged kidney
- Explain why the kidney is in that state of health.

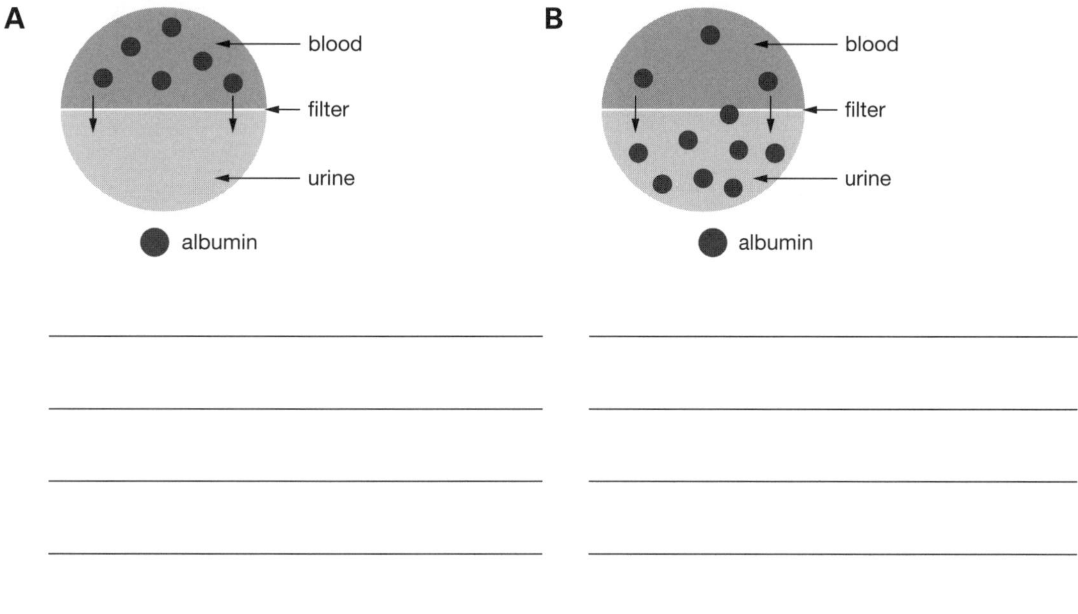

(b) Urine albumin levels are shown on the diagram below. Thirty is the critical level separating normal levels from high level. Numbers above 30 indicate poor kidney function.

On the diagram:

- Shade the section of normal kidney function in green.
- Shade the section of poor kidney function in red.
- Place a V on the section where some kidney disease symptoms will begin to appear.
- Place a W on the section where the patient is likely to need dialysis treatment.
- Place a X on the section where the patient may treat the kidney disease with medication.
- Place a Y where there is no albumin in the blood.
- Place a Z where kidney disease is serious but not critical enough for dialysis.

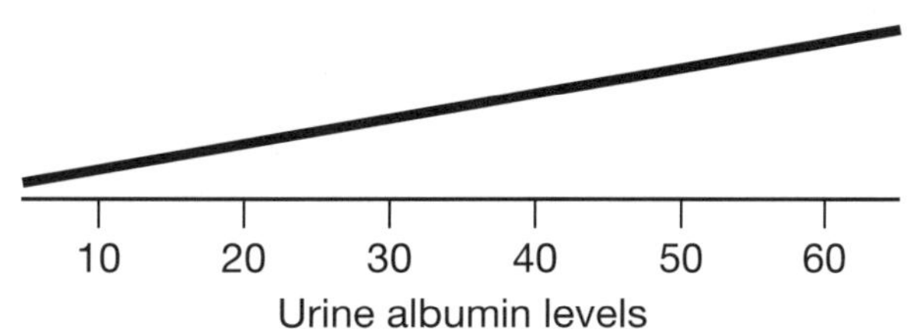

RATE MY UNDERSTANDING
Shade the face that shows your rating  

3.9 The skeleton

Science understanding

FOUNDATION	**STANDARD**	ADVANCED

1 Use the words in the box to label the different parts of the human skeleton.

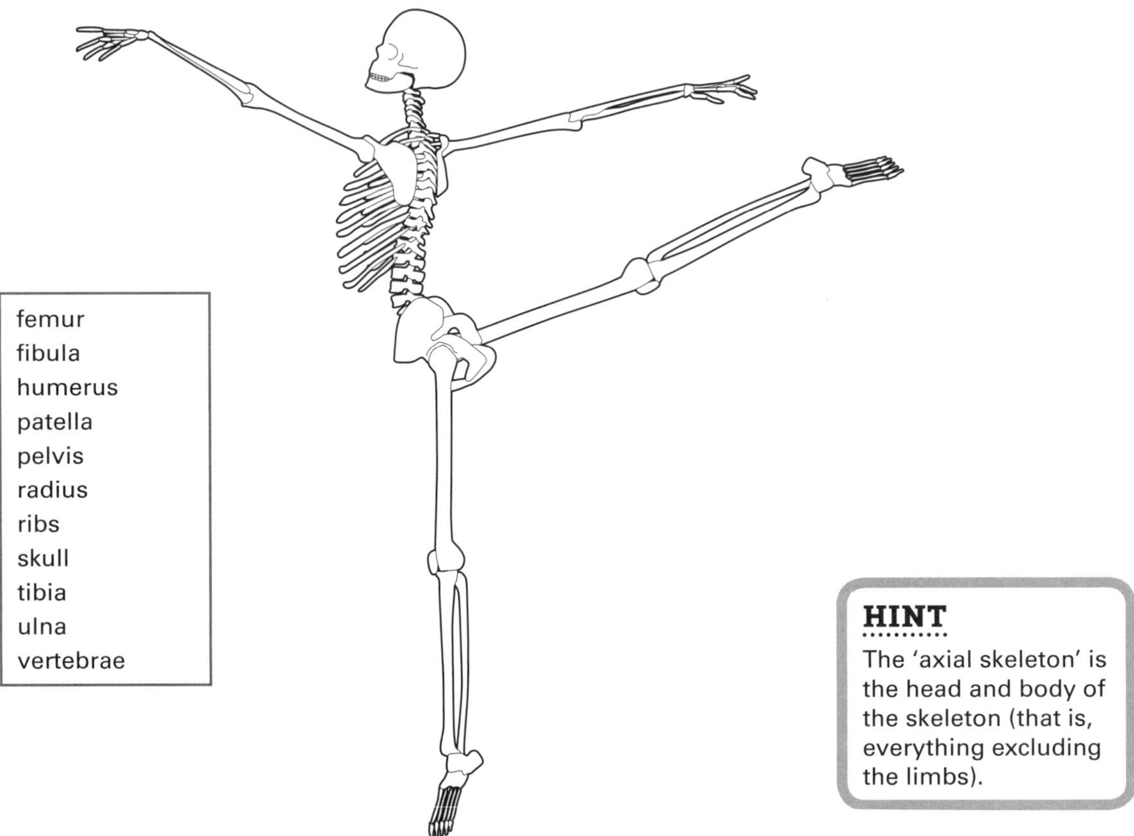

2 Colour the axial skeleton in red.

3 Label an example of each of the following joints and circle them in the colour indicated:

(a) hinge joint (circle in green)

(b) pivot joint (circle in blue).

(4) Explain how the joints of the skeleton allow the dancer to hold her right leg in the position shown in the diagram.

____

____

(5) Compare the range of movement of a hinge joint and a ball and socket joint. Provide examples of each type of joint.

____

____

____

RATE MY UNDERSTANDING
Shade the face that shows your rating

3.10 Spray on skin

Science as a human endeavour

FOUNDATION	STANDARD	**ADVANCED**

Professor Fiona Wood came into the public spotlight after the Bali bombings in 2002. 202 people were killed, including 88 Australians, and more than 200 were injured when suicide bombers attacked two nightclubs. Professor Wood was a plastic surgeon working at the Royal Perth Hospital at that time. She led a team of specialist health professionals to save 28 survivors, some of whom had up to 92 percent of their body burnt. These patients were susceptible to deadly infections and shock.

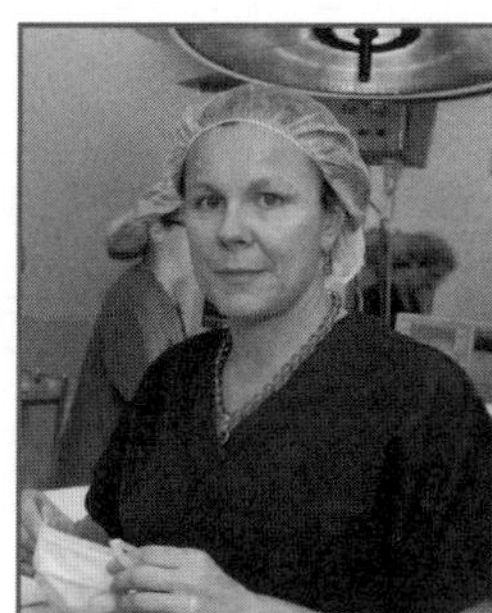

Professor Fiona Wood. Professor Wood was awarded the Order of Australia in 2003 and was Australian of the Year in 2005.

Professor Wood's early career in medicine was in plastic surgery and research. In the 1990s she developed an interest in burns medicine, especially in reducing the amount of scarring suffered by burns victims. Professor Wood worked with another scientist, Marie Stoner, to develop the technique of cultured sheets of skin which was an emerging technology being investigated in the United States. The two scientists moved from growing skin sheets to spraying on skin cells. As part of their work they were involved in conducting clinical trials to make sure procedures were safe and effective.

Professor Wood, through her work with burns patients and plastic surgery, has become internationally recognised for her patented invention of spray on skin. Before this invention, it took up to 21 days to culture enough cells to treat burns patients. Spray on skin has reduced this culture time to five days and this has in turn reduced the level of scarring. The faster a burn heals the less scarring there is. The technique is being continually developed and the goal is to eventually produce wound healing that leaves no scars.

The work of Professor Wood and her colleagues has gone a long way to improving the lives of burns patients and others who require skin grafts.

1 State how Fiona Wood was involved with the development of improved treatment for burns victims.

____

____

____

____

(2) Propose a reason why burns victims may have a high risk of deadly infections and delayed shock.

____

____

____

____

(3) Explain the advantages of spray on skin.

(4) Explain why scientists do clinical trials.

5 List the methods of healing burns that Professor Wood researched and helped develop.

Methods of treating burns victims

Over the last twenty years, there has been a great improvement in ways to aid the healing of burns victims. Three methods are described below.

Skin grafting	Cultured epithelial autograph (CEA)	Spray on skin
• Healthy skin is harvested from the victim. • The pieces of healthy skin are placed on the burns. • There is considerable scarring, loss of mobility and disfigurement after the burn has healed. • Patients may have long hospital stays.	• Healthy skin is harvested from the victim. • The healthy skin is grown into sheets in a laboratory. • It can take several weeks to grow enough cells to apply to the burn. • The skin sheets are applied onto the burns of the victim.	• Surgeons take a small piece of healthy skin from the victim. • The skin is cultured in a laboratory so more skin cells grow. • The process of growing enough cells takes about one week. • The cultured skin cells are sprayed onto the victim's burn. • There is little scarring after healing. • Patients have short hospital stays. • Healing process is complete after 6 months.

6 Complete the paragraph below using the information about methods of treating victims of burns. Correctly fill the spaces with words from the box below.

autograph	epithelial	faster	grafting	healthy	laboratory
long	more	quickly	scars	skin	spray on skin

All the treatments for burns involves taking small healthy ____________ pieces from the victim. Skin ____________ involves taking the piece of ____________ skin and placing it onto the burn. Then there is a ____________ waiting period for that skin to grow and heal the burn. The healed wound may have a lot of ____________ and skin may be tight and uncomfortable. Another treatment involves the healthy skin being cultured in a ____________ to grow a sheet of skin which is placed over the burn. This is called cultured ____________ autograph. The healed burn is not as scared because of the ____________ healing time, compared with skin grafting. Like skin grafting and cultured epithelial ____________, the ____________ method also involves the culture of healthy skin cells from the victim. The skin cells are cultured to grow ____________ cells and then made into a spray that is applied to the burn. Burns heal ____________ and there is little scarring or disfigurement.

3.11 Literacy review

Science understanding

FOUNDATION	STANDARD	ADVANCED

Recall your knowledge of human body systems by matching the key words on the left with their definitions on the right. Using a ruler, draw a line between the dots next to the matching terms or definitions. The line you draw should pass through one of the letters in the middle column. Reading down, the letters should spell out a key term relevant to this chapter.

antagonistic	E A	reactions that change food chemically
urine	S X	describes a pair of muscles that work in opposition to each other
excretion	C R	the tube that carries air from the nose and mouth into the chest cavity
chemical digestion	R	bony structure that holds body upright and protects organs
trachea	E T O	the material that has been filtered out of the blood by the kidneys
villi	O T N	cluster of sacs where gas exchange takes place
skeleton	S O	getting rid of the wastes the body has produced
aorta	M R	the artery that carries blood from the heart to the lungs
atrium	T Y Y	microscopic 'fingers' that greatly increase the surface area of the wall of the small intestine
alveoli	T M S	the main artery leaving the left ventricle of the heart
circulatory system	Y C	the system of the body that carries materials around the body, comprising the heart, blood vessels and blood
diaphragm	X O S	one of the chambers at the top of the heart that receives blood into the heart
tendons	T	the lower chambers of the heart that contract, pushing blood out of the heart
pulmonary artery	M E X	a sheet of muscle that separates the chest from the abdomen
ventricles	N Y M	elastic tissue that attaches muscle to bone

Key term: ______________________

RATE MY UNDERSTANDING
Shade the face that shows your rating

3.12 Thinking about my learning

1 Think back over your learning. Make a list of ten of the most important and interesting things you learnt on body systems. Use this list to create six questions. Use the who, what, where, when, why and how framework. Clearly write these six questions on the front of an envelope.

2 Rewrite each question on a sheet of paper, allowing room for answers. In a different colour, write detailed answers to each question. Place the answer sheet inside the envelope and seal the envelope.

3 Swap your envelope with a member of your class. Read your partner's questions and answer them in your notebook. Check your answers by opening the envelope and reading the answers supplied.

CHAPTER 4

Reproduction

4.1 Knowledge preview

Science understanding

FOUNDATION	STANDARD	ADVANCED

1 The following is a list of phrases which are associated with reproduction. Use the right-hand column of the table to indicate whether the phrase is associated with asexual reproduction or sexual reproduction. Write the correct type of reproduction in the space.

Phrases	Sexual or asexual reproduction?
(a) involves only one individual	
(b) produces offspring that are not identical	
(c) involves two individuals	
(d) occurs with bacteria	
(e) involves sperm and eggs	
(f) involves spores	
(g) involves splitting in two by fission	
(h) produces identical offspring	
(i) involves fertilisation	
(j) occurs with flowering plants	
(k) produces an embryo	
(l) called vegetative reproduction in plants	

2 The box below lists words associated with the reproductive systems of flowers and humans. Place the words in the Venn diagram that are just flower parts, just human parts or could be associated with either.

anther	gamete
ovary	oviduct
petal	pollen
prostate gland	sperm duct
stigma	style
testes	urethra
uterus	vagina

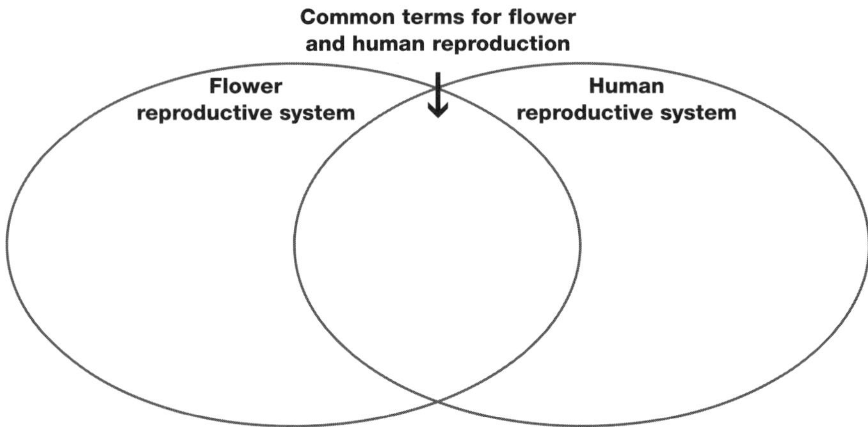

4.2 Flowers and pollination

Science understanding

FOUNDATION	**STANDARD**	ADVANCED

Flowers can have many different shapes, sizes and colours. Careful study of different flowers has led biologists to conclude that these differences are related to how the plants reproduce. Table 4.2.1 is a summary of how the main features of cross-pollinating flowers depend on the way in which they are pollinated.

Main features of cross-pollinating flowers			
Method of pollination	**Flower structure/colour/size**	**Anther/stamens**	**Stigma/style**
Wind	often small but with many flowers in one head, often no petals, not brightly coloured, no nectar, no scent	long stamens with large anthers exposed	long style with exposed stigma; stigma has large surface area—often look like brushes
Insect	usually small, some with many flowers in one head, brightly coloured petals especially blues and yellows, small amounts of nectar, strong scent, often strongly marked with 'landing guides'	often short stamens and small anthers, close to nectar source in most flowers, sticky pollen	short style, small stigma close to nectar source
Bird	large strong flowers, some have petals but many don't, lots of nectar, often red	often long, strong stamens and large anthers, sited a long way from the nectar source	long style, smallish stigma, sited a long way from the nectar source
Mammal	large strong flower heads, often not brightly coloured and often hidden in plant, much nectar produced at night	strong and rigid	strong and rigid

Table 4.2.1

Some characteristics of Australian animals that pollinate flowers have also been studied, and are shown in Table 4.2.2.

Characteristics of Australian animals that pollinate flowers	
Animal	**Characteristics**
insects	poor eyesight, low intelligence, good sense of smell, small bodies
birds	good eyesight, intelligent, most active in daylight (diurnal), poor sense of smell, large bodies
mammals	good eyesight, intelligent, most active at night (nocturnal), good sense of smell, large bodies

Table 4.2.2

1. A flower's features are related to the way its pollen is carried. Explain why bird-pollinated flowers would be larger and stronger and have more nectar than insect-pollinated flowers.

____

____

2. Suggest reasons why wind-pollinated flowers would have large brush-like stigmas and large anthers with a lot of pollen.

____

____

4.2 Flowers and pollination

(3) Explain why mammal-pollinated flowers in Australia lack colour, have a strong scent, and are often hidden away, for example *Grevillea leucopteris*.

(4) Which method of pollination occurs for each flower shown below? Justify your answer.

(a) Kangaroo paw

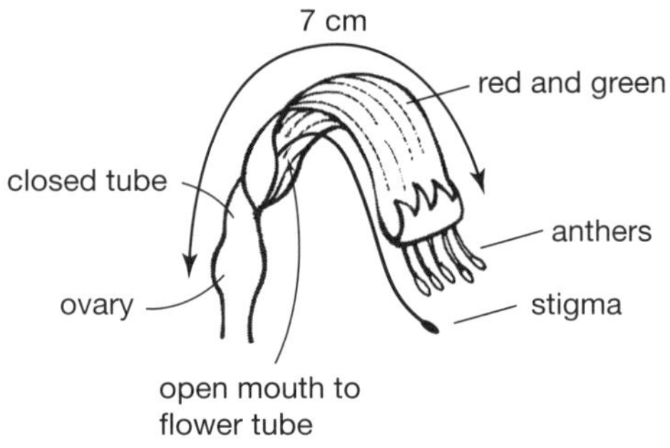

(b) Starflower

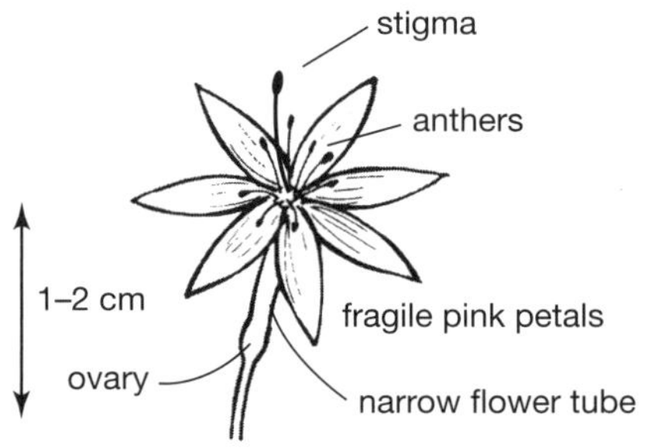

(c) Veldt grass

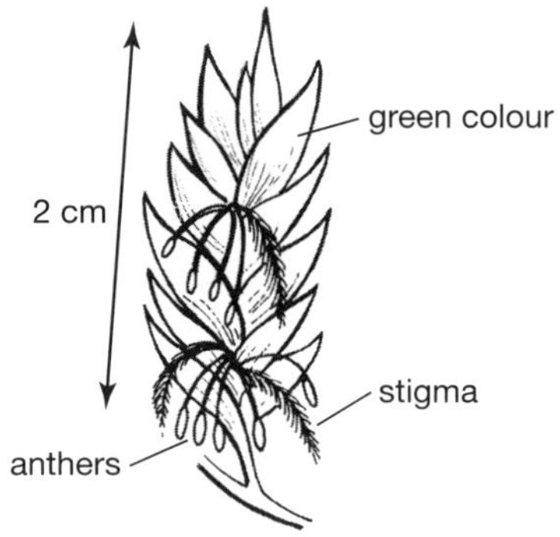

(d) Westringia

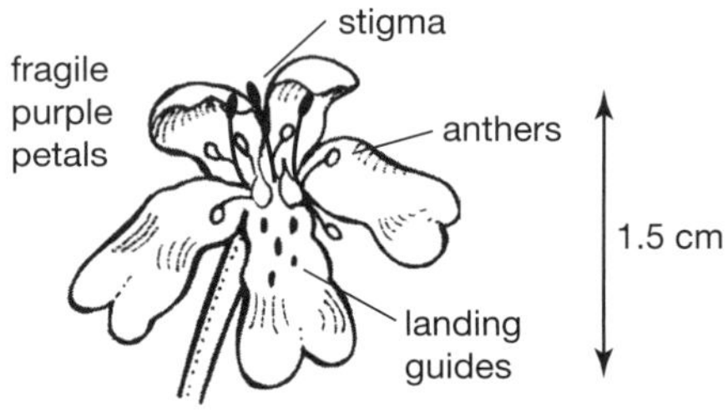

(e) Nodding banksia

flower head (about 10 cm long)

brown colour, hanging near surface and scented

ovary

anthers

stigma

one flower

5. Some orchids have a flower that resembles a female wasp. The male wasp is tricked into trying to mate with the flower, as shown in Figure 4.2.1. A packet of pollen sticks to the male wasp's abdomen, when he attempts to mate. When the wasp flies to another orchid and again attempts to mate, this pushes the pollen into the flower and pollinates it. Wasps have fairly poor vision and cannot see more than a few metres. Explain how the wasp finds the flower.

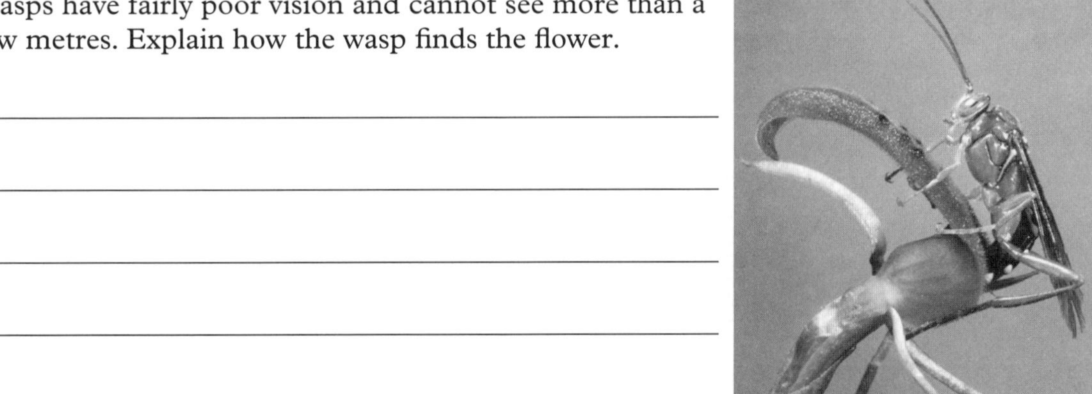

Figure 4.2.1 Male wasp pollinating a flower

6. Design your own original and unique flower.
   - Give your flower a name.
   - Label all the parts of the flower.
   - Describe the method by which your flower will be pollinated.
   - Explain what characteristics of the flower will enable it to be pollinated this way.

abdomen (*n*) the part of the body containing the stomach

vision (*n*) eyesight

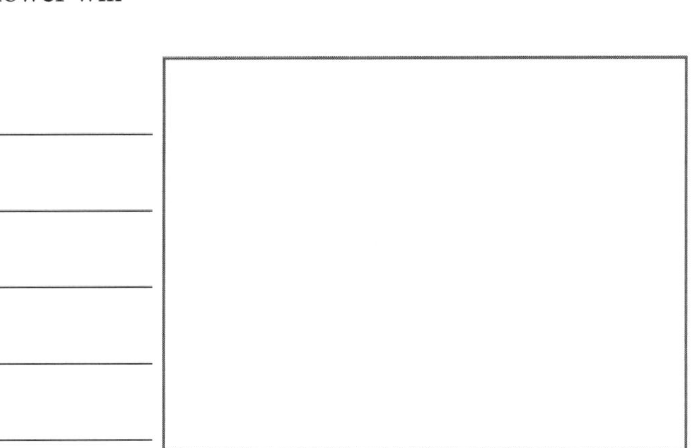

RATE MY UNDERSTANDING

Shade the face that shows your rating

4.3 Aphid life cycles

Science understanding

FOUNDATION | **STANDARD** | ADVANCED

Aphids have amazing life cycles. They can reproduce by two different methods, both sexually and asexually (Figure 4.3.1). Aphids reproduce asexually by parthenogenesis. In this process, females produce eggs that can grow into young without being fertilised by a male sex cell. The eggs develop inside the female body and she gives birth to baby aphids that look like small adults.

The method of reproduction aphids use depends on the weather conditions where they live. The main factor that affects them is the temperature, which determines the sex of the young.

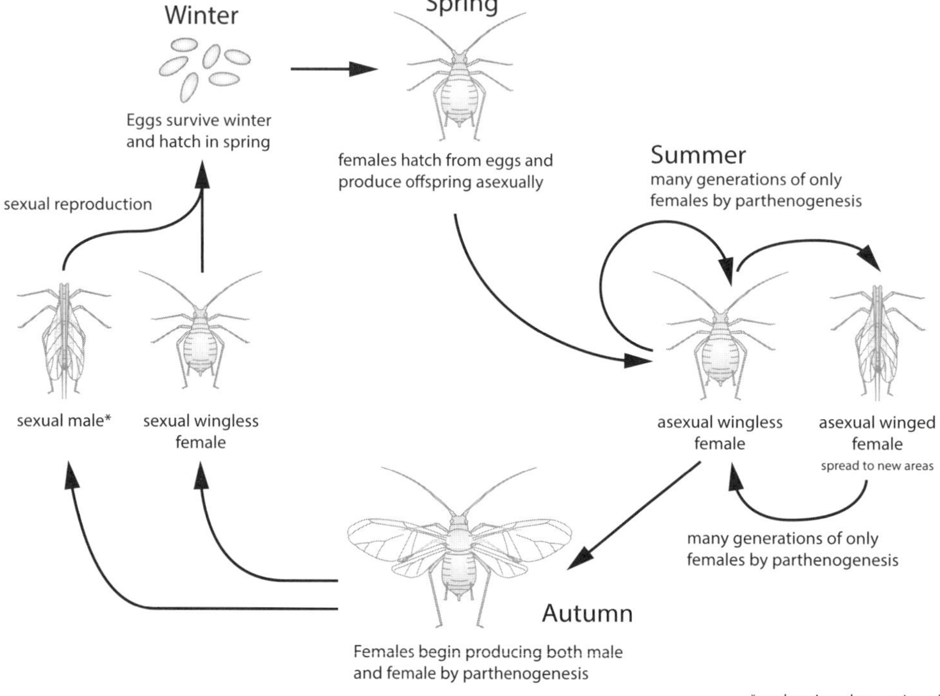

Figure 4.3.1 The life cycle of aphids involves both sexual and asexual reproduction

Asexual stage

In warm regions, like most of Australia in summer, the young produced by parthenogenesis are nearly always female. This is shown in Figure 4.3.1 in the section labelled 'Summer'. These females produce many generations of only female offspring. Most of the young produced by asexual reproduction in summer do not have wings. However, some females are born with wings.

Figure 4.3.2 An aphid giving birth to a baby by parthenogenesis.

The females with wings spread the population to many plants, which provides a large food supply. The large food supply and warm temperatures help the aphids to reproduce very fast. This rapid reproduction occurs mainly because they are reproducing by parthenogenesis. Parthenogenesis produces offspring much faster than sexual reproduction, mainly because females do not have to find a male to breed.

4.3 Aphid life cycles

Sexual stage

Aphids can reproduce sexually as well. This process starts in the autumn in cold regions where snow and ice occur. This is shown in Figure 4.3.1 in the 'Autumn' section. Females begin to produce both males and females by parthenogenesis. Some males have wings and can fly to plants where there are females. Other males do not have wings but can find females on plants nearby. The males mate with females and fertilised eggs are produced that are laid by the females onto a leaf. The eggs are able to resist low temperatures and so survive the winter, whereas most of the adults do not. The eggs hatch in spring into females only. These females then reproduce by parthenogenesis and the process repeats itself.

1 Identify the main environmental factor that determines whether aphids reproduce sexually or asexually.

____

2 Name the process by which aphids reproduce asexually.

____

3 Describe the process that you named in question 2.

____

____

4 In very cold places, as the temperature drops, what changes occur in the offspring produced by parthenogenesis?

____

____

(5) For very cold places, predict what life stage the aphids would be in during the winter.

____

(6) Explain why the aphids that are found in most places in Australia are nearly always female and reproduce by parthenogenesis.

____

____

4.4 Human reproductive system

Science understanding

FOUNDATION	STANDARD	ADVANCED

1 Select words from the box below to name the parts of the female reproductive system in Figure 4.4.1.

cervix
ovary
oviduct
uterus
vagina

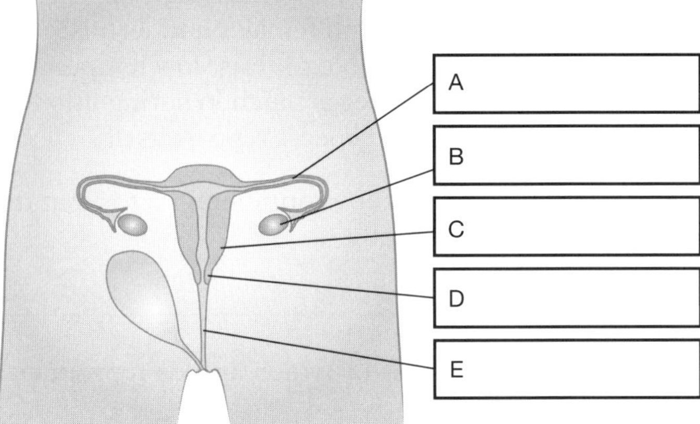

Figure 4.4.1 Female reproductive system

2 Describe the function of each of the following parts of the female reproductive system.

(a) ovary ______________________

(b) uterus ______________________

(c) oviduct ______________________

(d) vagina ______________________

3 Select words from the box below to name the parts of the male reproductive system in Figure 4.4.2.

Cowper's glands
epididymis
penis
prostate gland
seminal vesicle
sperm duct (vas deferens)
testicle (testis)
urethra

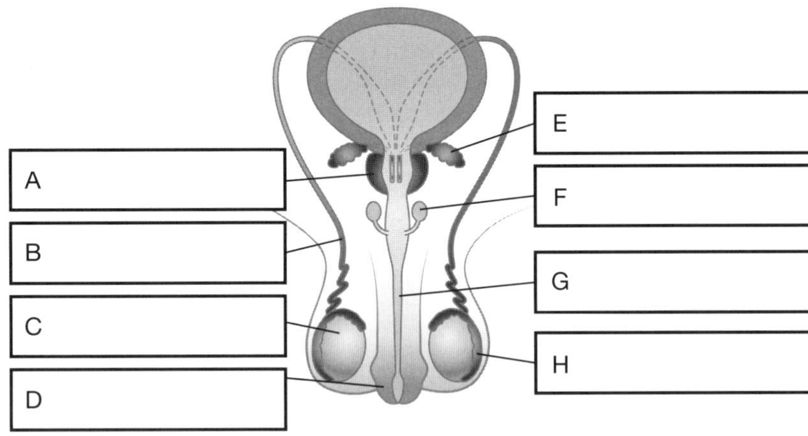

Figure 4.4.2 Male reproductive system

4 Describe the function of each of the following parts of the male reproductive system.

(a) testes ______________________

(b) epididymis ______________________

(c) sperm duct ______________________

(d) penis ______________________

RATE MY UNDERSTANDING
Shade the face that shows your rating   

4.5 Puberty and growth

Science understanding

FOUNDATION | **STANDARD** | ADVANCED

Puberty is the time in a young person's life when they start to develop physically and emotionally. Their bodies begin to change as they move from childhood into adolescence. Girls and boys grow at different rates and have growth spurts at different ages as shown in Table 4.5.1.

Table 4.5.1 The average heights of boys and girls at different ages

Age (years)	Male height (cm)	Female height (cm)
Birth	50.5	50.2
2	87.5	86.6
4	103.4	103.2
6	117.5	115.9
8	130.0	128.0
10	140.3	138.6
12	149.6	151.9
13	155.0	157.1
14	162.7	159.6
16	171.6	162.2
18	174.5	162.5

1. Use the axes on the graph provided to construct line graphs showing the heights of males and females. Add a suitable scale to both axes. Include a key and title.

2. What are the ages when girls grow faster than boys?

____

____

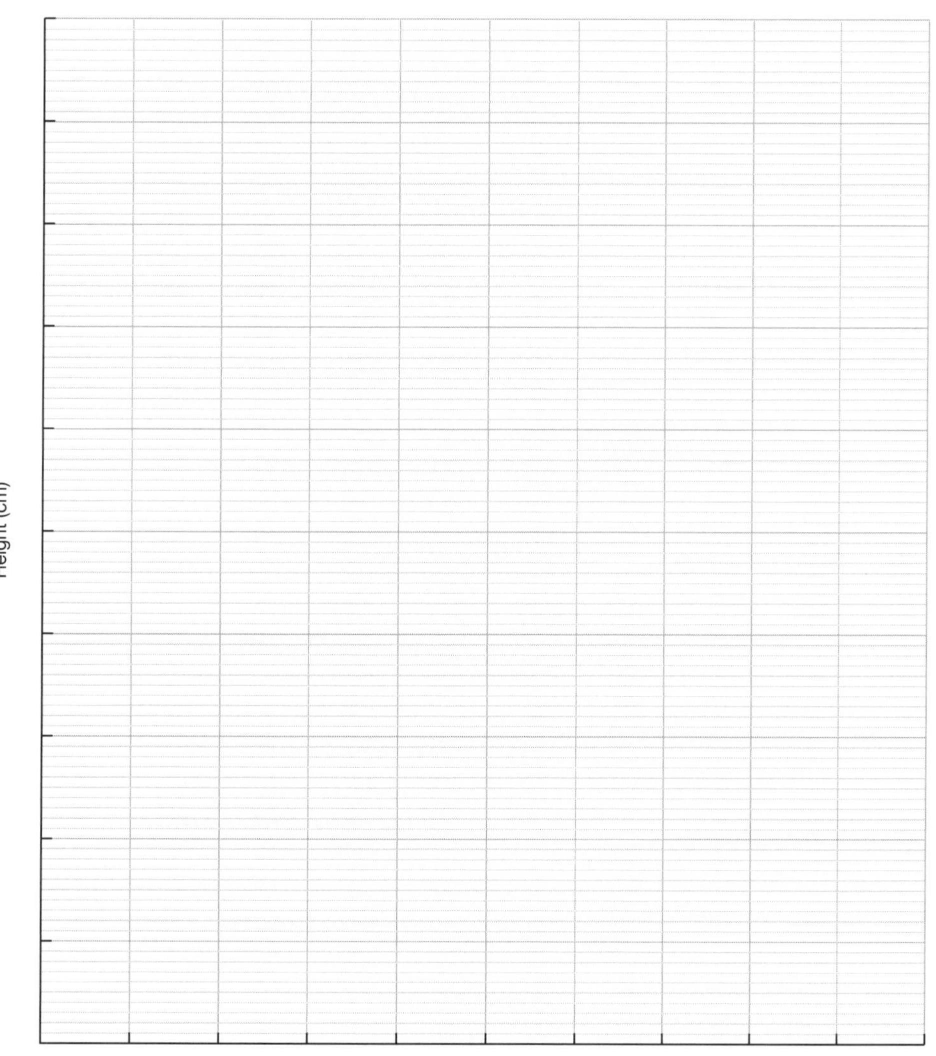

4.5 Puberty and growth

(3) Compare the growth patterns of boys and girls up to age 10.

HINT
Compare means to find what is similar and what is different

(4) Compare the growth patterns of boys and girls between ages 10 and 16.

5 Many people say 'girls mature faster than boys.' Use your graph as evidence to justify this claim.

6 Suggest reasons why growth spurts occur.

RATE MY UNDERSTANDING
Shade the face that shows your rating

4.6 Pregnancy

Science understanding

FOUNDATION	STANDARD	ADVANCED

1 Figure 4.6.1 shows the sequence of events from ovulation to implantation, beginning at text box labelled 1. In the blank boxes briefly describe the events at each stage. Include the names of any important structures.

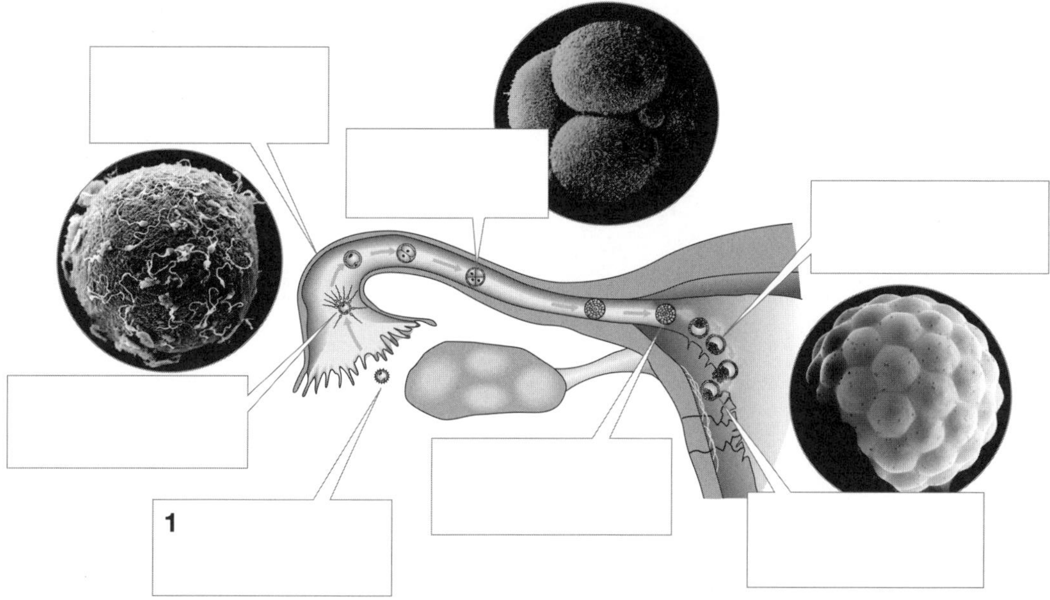

Figure 4.6.1 Ovulation to implantation

2 Figure 4.6.2 shows structures important in prenatal development.

(a) Name the structures numbered 1 to 8. Write the name next to the number.

(b) Describe the function of structure number 5.

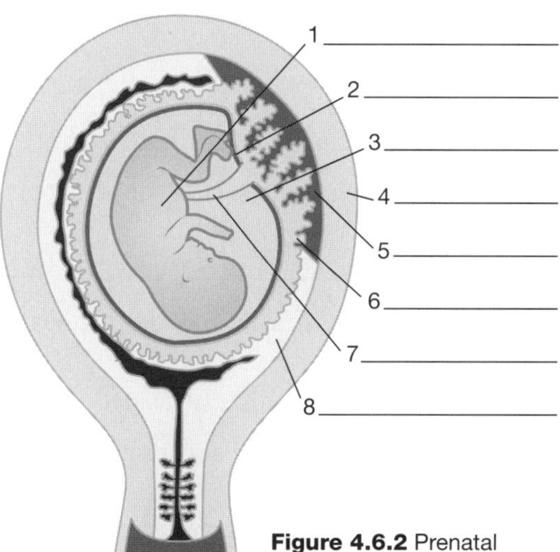

Figure 4.6.2 Prenatal development

(c) Describe the function of structure number 7.

RATE MY UNDERSTANDING Shade the face that shows your rating			☹

4.7 Reproduction, gestation and size

Science inquiry skills

FOUNDATION | STANDARD | **ADVANCED**

Processing & Analysing | Evaluating

The following chart compares the gestation period, lifespan and the size of various animals. Look at the chart information carefully and then answer the questions that follow.

> **lifespan** (*n*) the length of time that a person or animal lives

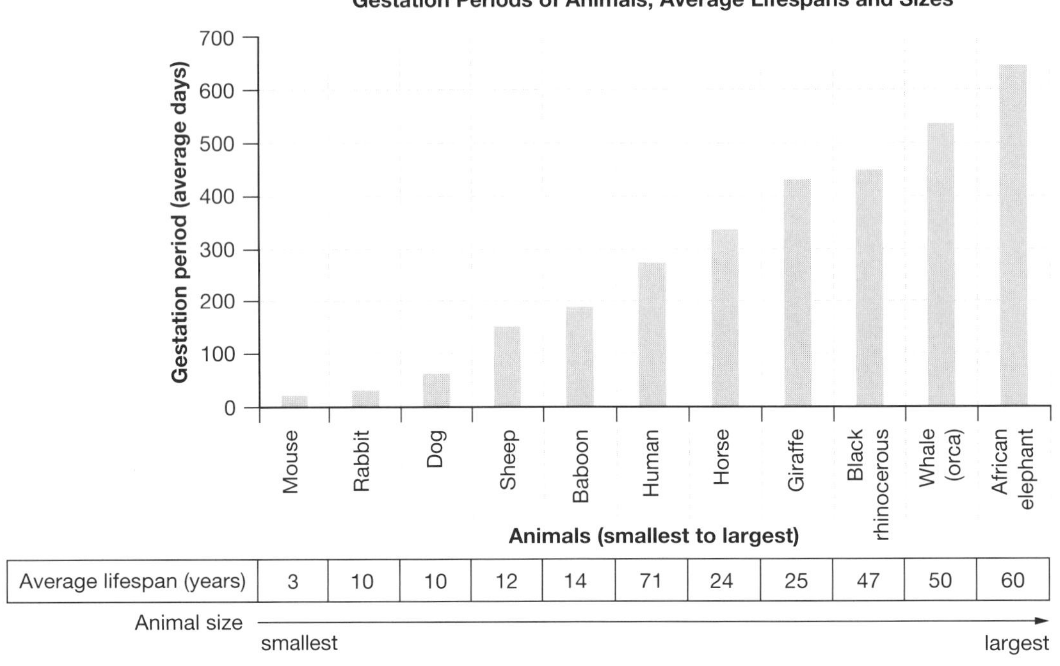

Average lifespan (years)	3	10	10	12	14	71	24	25	47	50	60

1 Explain what is meant by gestation period.

____

____

2 Which animal has the shortest gestation period?

____

3 Which animal has the longest gestation period?

____

4 Complete the sentences below by selecting from the words in the box and writing them in the spaces.

48	266	baboon	elephant	longer	shorter

Dogs have a ______________ gestation period than giraffes. There is a difference of about 620 gestation days between a mouse and a/an ______________. The whale has a longer lifespan than the ______________. Humans have a gestation period of ______________ days. On average, sheep live ______________ years less than African elephants.

5. Compare the gestation period and lifespan of humans and whales.

6. What overall conclusion can you draw from the data when you compare:

(a) gestation period with the size of the animal

(b) gestation period with life span.

7. For our body size, humans have a long gestation period. For example, the mass of a human is only about 20% of the mass of a horse, but humans have a gestation period that is about 80% that of a horse. Propose a biological reason why humans have a long gestation time for our body size. (Hint: What is the largest human body part at birth?)

8. Sea lions have an average gestation period of 350 days. Propose what the lifespan of a sea lion could be. Explain how you reached this conclusion.

RATE MY UNDERSTANDING
Shade the face that shows your rating

4.8 Fertility and IVF

Science as a human endeavour

FOUNDATION	STANDARD	**ADVANCED**

Some couples find that they cannot conceive children naturally. One common reason for infertility in women is blocked oviducts due to infection. Ova are produced but cannot pass down into the uterus. For men, a common condition is swollen veins inside their testes. The swollen veins reduce the amount of blood flowing to the testes causing overheating and insufficient nutrients for proper functioning of the testes. Lifestyle can also affect fertility. For example, being grossly overweight or experiencing excessive stress can reduce fertility in both males and females.

conceive (*v*) to become pregnant
fertilisation (*n*) the fusion of an egg and sperm to make a baby
grossly (*adj*) extremely, very
infertility (*n*) not being able to have children

In vitro fertilisation (IVF)

For couples where natural fertilisation cannot occur, treatment exists that makes it possible for women to have children. The process is known as in vitro fertilisation (IVF). Sperm and eggs are collected from a man and a woman and are mixed in a dish outside the woman's body (Figure 4.8.1). As long as the sperm and egg are healthy, fertilisation can occur. After a few days, when the embryo has reached the blastocyst stage, it is placed into the uterus of the female. The embryo then implants naturally in the lining of the uterus. When this happens, the woman is considered pregnant.

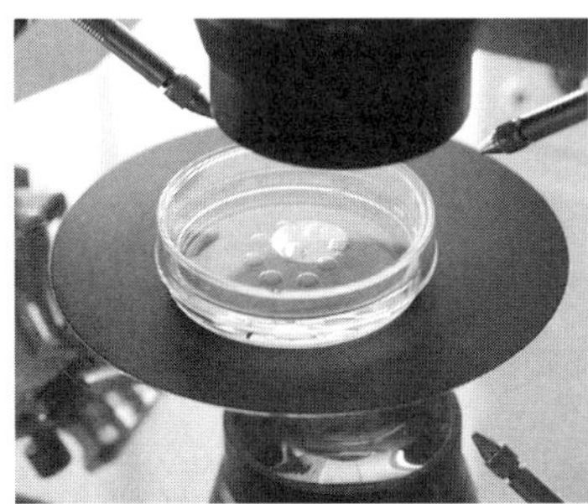

Figure 4.8.1 IVF

IVF has been used successfully since 1978 when the world's first IVF child, Louise Brown, was born in the United Kingdom. In 2007 Louise had a child of her own—a son who was conceived naturally.

1 Conduct some research to find the meaning of the term 'in vitro fertilisation' and explain where this originated.

____

____

2 **(a)** Use the flow chart to demonstrate the sequence of events in the IVF process.

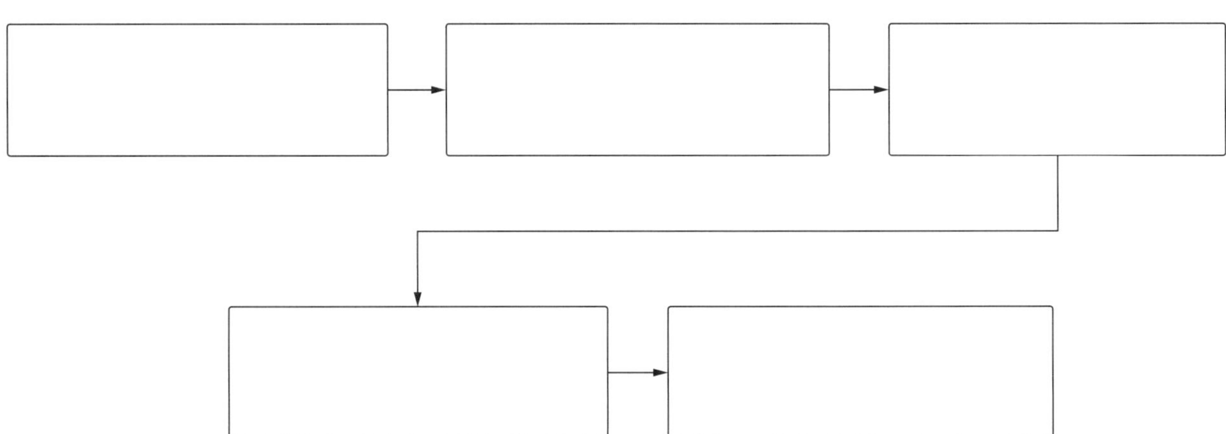

(b) Colour in the boxes of the flow chart where there are differences from natural reproduction.

(3) Complete the PMI chart (Plus, Minus, Interesting) on IVF treatment on page 65. Consider the advantages and disadvantages. You may wish to conduct some additional research.

4.8 Fertility and IVF

IVF treatment		
Plus	**Minus**	**Interesting**

4 The graph shows assisted reproduction data for Australia and New Zealand in 2014. One column shows the percentage of women who completed an initiated cycle assisted reproduction IVF and became clinically pregnant. The other column shows the percentage of those women who gave birth to a baby.

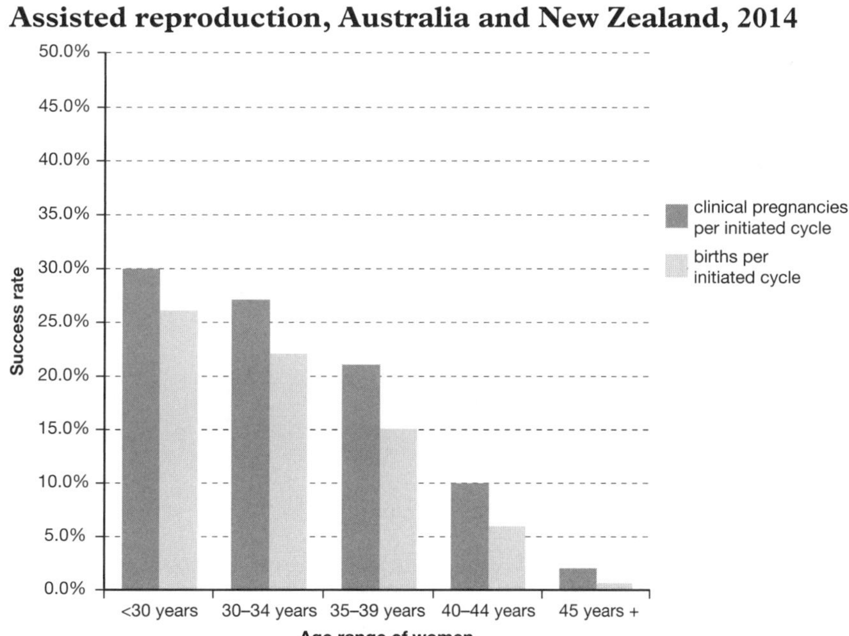

Indicate whether the following statements are true or false by writing the correct response in the space provided.

Statements	True/False
In the 40–44-year-old age group, 14% of women who completed an initiated cycle of assisted reproduction became pregnant.	
There was a greater success rate in becoming pregnant in the 30–34 age group than 35–39 age group.	
Forty percent of the women in the 30–34 age group gave birth.	
In the over-45 age group, all the pregnancies resulted in babies being born.	
The chance of IVF success is reduced as the mother's age increases.	
Twenty-five percent of the women aged 35–39 years gave birth.	
In the 30–34 age group 5% of the pregnancies were unsuccessful, because they did not result in a birth.	

RATE MY UNDERSTANDING
Shade the face that shows your rating

4.9 Literacy review

Science understanding

FOUNDATION | STANDARD | ADVANCED

1 Recall key terms and their meanings by drawing a line to match each statement in the left-hand column with the correct term in the right-hand column.

Statement	Term
a very early stage of human development before becoming a foetus	foetus
the joining of gametes	ovulation
tubes down which the egg passes and where fertilisation occurs	placenta
stage of development when most of the major organs and systems are present	germination
asexual reproduction by splitting in two	embryo
when the young growing plant sprouts out of the seed	amniotic fluid
the egg bursting out of the follicle	stigma
membranes of the fetus and mother that are in contact, allowing exchange of nutrients and oxygen for wastes	oviducts (fallopian tubes)
new individual grows from one parent's body and does not involve joining of gametes	fertilisation
flower part that receives the pollen	asexual reproduction
fluid around the fetus acting as a shock absorber and keeping the foetus at a constant temperature	fission

RATE MY UNDERSTANDING
Shade the face that shows your rating

4.10 Thinking about my learning

1 The following diagram plots your biological journey—the stages and processes your life has gone through to reach your present stage. Use the words in the box below to complete your biological journey on the flow chart provided. A circle is a life stage and a rectangle is a biological process. Use the space provided to write what you think will happen to you next.

baby	birth	child	egg	embryo	fertilisation
foetus	puberty	sperm	teenager	zygote	

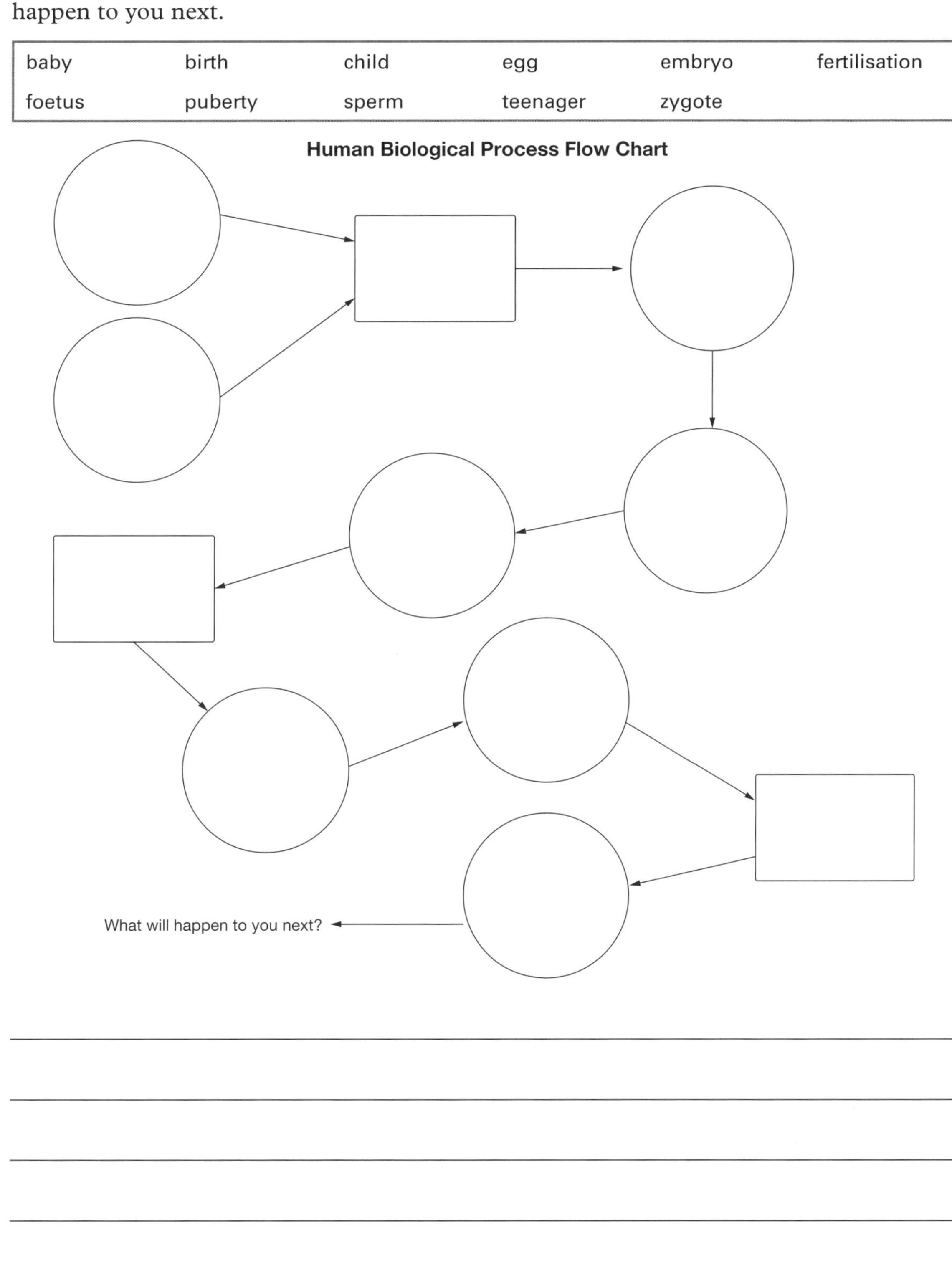

CHAPTER 5 Energy

5.1 Knowledge preview

Science understanding

FOUNDATION | STANDARD | ADVANCED

Energy comes in many forms. You may be familiar with some forms of energy but not with others.

1 **(a)** Consider the nine forms of energy listed in the box below and place them in the table according to how much you feel you know about them. You do not have to write words in every column.

chemical	elastic potential	electrical	gravitational potential	heat	kinetic	light	nuclear	sound

I am fairly sure I know what this is	I have some idea what this is	I do not know about this form of energy

(b) For each word that you have written in the first column, write a short sentence which shows what you know about that form of energy.

2 One of the things you will learn about in this chapter on energy is energy transformations. Energy is not the only thing that can be transformed.

(a) Explain what transformation means.

(b) Write a sentence describing a transformation which you have seen.

5.2 Forms of energy

Science understanding

FOUNDATION	STANDARD	ADVANCED

1. There are many different forms of energy. Select words from the box below to classify which energy is present in each of the stated situations. The first one has been completed for you.

chemical	elastic potential	electrical	gravitational potential	heat	kinetic	light	nuclear	sound

Situation	Types of energy present
(a) racing car starts a race	kinetic, heat, sound, chemical, electrical
(b) rubber ball warms up in the sun	
(c) hot air balloon sails above some clouds	
(d) springs on a trampoline are stretched before it bounces upwards	
(e) petrol is put into a car	
(f) desk lamp shines brightly	
(g) candle burns	
(h) ball rolls down a hill	
(i) boy brushes his teeth	
(j) cat climbs up a tree	

2. Kinetic energy is the energy of a moving object. Potential energy is the energy stored in an object. Classify each example below as having either kinetic or potential energy.

(a) slingshot about to fire ____________________

(b) ball at the highest point of a bounce ____________________

(c) swimmer about to dive from a high platform ____________________

(d) swimmer hitting the water from a dive ____________________

(e) teenager skating along a footpath ____________________

(f) hamburger with the lot sitting on a plate ____________________

(g) stone rolling along a road ____________________

(h) new packet of AA batteries ____________________

(i) bowl of cereal with milk ____________________

(j) leaf on a tree ____________________

RATE MY UNDERSTANDING
Shade the face that shows your rating

5.3 Energy changes

Science understanding

FOUNDATION | **STANDARD** | ADVANCED

Energy makes things happen. When something happens, energy may be passed (transferred) from one object to another. Energy transfer happens when you hit a tennis ball. Some of the kinetic energy of the racquet is transferred to the ball. Energy can also be changed (transformed) into another type of energy. When you hit the tennis ball, chemical energy from food that you ate is transformed into kinetic energy in your arm.

1 For each example below, the source of the energy is given. State the receiver(s) of this energy and identify whether energy was transferred or transformed in the process. The first one has been completed for you.

> **transfer** (*v*) to move from one place to another
> **transform** (*v*) to change

Example	Source of energy	Receiver of energy	Is energy transferred or transformed?
(a) Tom runs in a race.	chemical energy (food)	muscles in Tom's arms and legs	transformed (into kinetic energy)
(b) A shirt hanging on a washing line dries in the sun.	heat energy (the Sun)		
(c) An aeroplane takes off.	chemical energy (fuel)		
(d) A golf club hits a golf ball.	kinetic energy (the club)		
(e) A cat warms up by an open fire.	heat energy (the fire)		
(f) A ceiling fan is switched on.	electrical energy		

2 An energy flow diagram is a way of showing the energy changes that happen. Construct an energy flow diagram for the following energy changes. The first one has been completed for you.

> **carve** (*v*) to cut, slice
> **scuttle** (*v*) to move quickly
> **wind-up beetle** (*n*) a mechanical toy

(a) A petrol lawn mower cuts some grass.

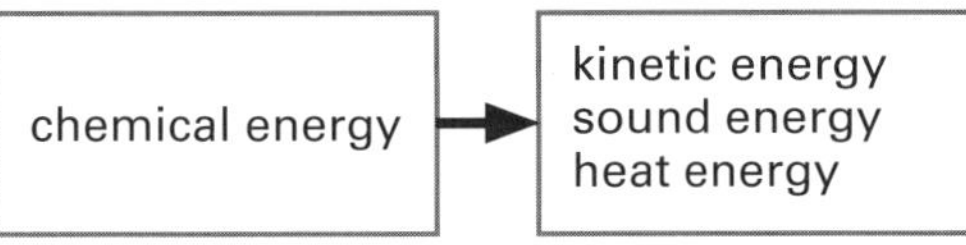

(b) A solar cell is used to operate an outside light.

(c) An electric knife is used to carve a roast.

(d) A wind-up beetle is released and scuttles across the floor.

RATE MY UNDERSTANDING
Shade the face that shows your rating

5.4 Energy efficiency

Science understanding

FOUNDATION | **STANDARD** | ADVANCED

Many household devices convert electrical energy into other forms.

(1) Identify the types of energy, both useful and wasted, that result from transformations by the following devices. The first example has been done for you.

Device	Useful energy produced	Wasted energy produced
(a) electric mixer	kinetic	heat, sound
(b) chain saw		
(c) radio		
(d) treadmill		
(e) desk lamp		

2 The following devices convert the electrical energy supplied into specific amounts of other forms of energy. Calculate the missing Joule (J) values to complete the following energy conversions. Remember that energy is conserved, not lost.

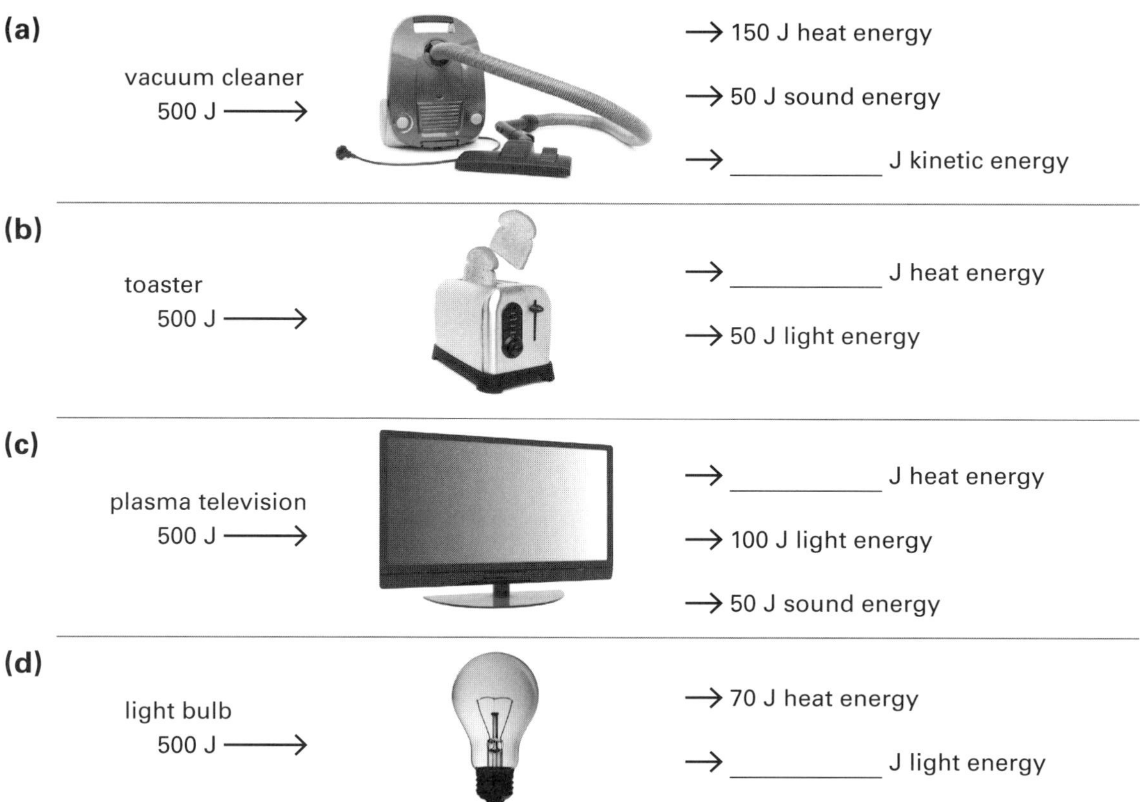

3 The efficiency of the devices in the previous question can be calculated using the equation below. The useful energy output is the energy the item is designed to make and not the waste energy that may also be made. The energy efficiency is expressed as a percentage.

$$\frac{\text{useful energy output (J)}}{\text{energy input (J)}} \times 100$$

Worked example:

$$\text{Percentage efficiency} = \frac{\text{useful light energy output (J)}}{\text{total energy input to light bulb (J)}}$$

$$= \frac{430\text{ J}}{500\text{ J}} \times 100 = 86\%$$

(A light bulb is designed to give light not heat, so the useful energy is light.)

(a) Use the equation and data in question 2(b) to calculate the efficiency of the toaster.

____

____

(b) Assuming that light and sound are the useful forms of energy from a television, calculate the efficiency of the plasma television using data from question 2(c).

____

____

(c) State whether the toaster or the television is more energy efficient.

____

RATE MY UNDERSTANDING
Shade the face that shows your rating

5.5 Conserving energy

Science inquiry skills

FOUNDATION	STANDARD	ADVANCED

If we can reduce our energy use, then we will reduce greenhouse gas emissions. Some ways of saving energy are very simple, such as turning off an appliance at the power point when it is not being used. The diagram below gives some tips on how you can save energy around your home.

appliance (*n*) a machine in the home in the home that uses electricity
emission (*n*) something that comes out
greenhouse gas (*n*) a gas that causes the Earth's atmosphere to heat up
tips (*n*) suggestions, useful ideas

(1) Propose as many additional ways to save energy as you can and add these to the diagram.

Bedroom

- close curtains in the evening to save on heating

Bathroom

- have short showers

Kitchen

- thaw frozen foods before cooking
- only run a dishwasher when it is full
- use lids when heating food in saucepans

Study

- turn computer screen off if not being used

Garage

- try to walk, ride or catch public transport so that your family uses the car less

Lounge room

- close blinds, curtains and windows on a hot day
- don't use heaters overnight or when no one is home

Laundry

- try to dry clothes on the washing line, or on a rack above a heating duct or near a heater (if safe)
- wash clothes in cold water

RATE MY UNDERSTANDING
Shade the face that shows your rating

5.6 Energy efficiency ratings

Science as a human endeavour

FOUNDATION | **STANDARD** | ADVANCED

The Australian Government passed the Greenhouse and Energy Minimum Standards Act (GEMS) in October 2012. This Act created an energy efficiency framework, setting minimum energy efficiency standards and requiring energy rating labels for many products.

The energy rating label gives consumers information about the energy performance of products. Customers can compare similar models of a product by looking at the star ratings and yearly energy use.

1 Conduct research to find out which products must have an energy rating label in Australia, then list them.

____

____

(2) There are two types of Energy rating labels—the 6-star label and 10-star label for appliances with efficiencies higher than 6 (Figure 5.6.1). The more stars on the label, the greater the efficiency. Labels also provide the amount of electricity the appliance uses in a year. The lower the kilowatt-hours (kWh) the less electricity used, the less it costs to run an appliance and the greater the efficiency.

Figure 5.6.1 Energy rating labels

C

SUPER EFFICIENCY RATING

The more stars the more energy efficient

ENERGY RATING

Energy consumption

kWh per year

Compare models at www.energyrating.gov.au

Figure 5.6.2

(a) Compare the two Energy rating labels in Figure 5.6.1. They provide information on two different models of washing machines. Which washing machine, A or B, is more energy efficient? Give reasons for your answer.

____

____

(b) A third washing machine is also being considered for purchase by a customer. It has an 8-star rating and uses 170 kWh of electricity per year. Complete the label in Figure 5.6.2 to show the energy data.

(c) Compare all three washing machines. Which is the most energy efficient and why?

____

____

RATE MY UNDERSTANDING
Shade the face that shows your rating

5.7 Musical Instruments

Science understanding

FOUNDATION	STANDARD	ADVANCED

Sound is produced by an initial vibration. All musical instruments also rely upon an initial vibration to produce sound. The body of the instrument acts as a sound board to amplify the sound produced. In the case of woodwind instruments such as saxophones and clarinets, a thin piece of wood called a reed vibrates inside a mouthpiece. Brass instruments, such as the trumpet, trombone and tuba also have a mouthpiece but do not have a reed. Instead, the buzzing lips of the musician sets up the initial vibration. If you have ever looked inside a piano you might have seen that when a key is struck, a padded hammer strikes a stretched string to produce vibration of the correct pitch.

1. Look at each of the following musical instruments. Identify what causes the initial vibration and write your response next to each instrument. You may find it helpful to conduct some research on each instrument.

Musical instruments and the part that vibrates first when played		
(a) piano  What vibrates?	**(b)** guitar  What vibrates?	**(c)** sousaphone  What vibrates?
(d) violin  What vibrates?	**(e)** xylophone  What vibrates?	**(f)** drums  What vibrates?
(g) saxophone 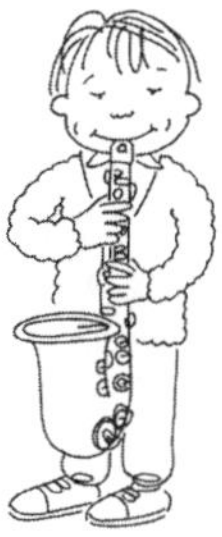 What vibrates?	**(h)** tuba  What vibrates?	**(i)** flute  What vibrates?

RATE MY UNDERSTANDING
Shade the face that shows your rating

5.8 Reflection or refraction?

Science understanding

FOUNDATION	STANDARD	ADVANCED

1 Classify each of the following situations as examples of either reflection or refraction. Circle the most correct response for each situation.

Situation		Situation	
(a)	reflection refraction	(b)	reflection refraction
(c)	reflection refraction	(d)	reflection refraction
(e)	reflection refraction	(f)	reflection refraction
(g)	reflection refraction	(h)	reflection refraction
(i)	reflection refraction	(j)	reflection refraction

RATE MY UNDERSTANDING
Shade the face that shows your rating

5.9 Literacy review

Science understanding

FOUNDATION	STANDARD	ADVANCED

1 Read each clue in the left column and match it to the correct term in the right column, by drawing a line between them.

Clue
We measure energy using this unit.
This is energy of movement.
This energy warms you up.
This energy enables us to see.
This energy is caused by vibrations.
This type of energy powers a television.
Energy is passed from one object to another but is not changed.
Stored energy due to height above the ground is called ______.
The stored energy found in food and fuel.
Energy stored in a stretched rubber band.
Energy stored inside the particles that make up matter.
A measure of the proportion of useful energy that is produced by a device.
A law that states that energy can never be created or destroyed is called the law of ______ of energy.
A label showing a number of stars that is used to compare energy efficiency of appliances is called the energy ______ label.
Energy is changed from one form into another.
Method of heat transfer that occurs between solids in contact.
Method of heat transfer that occurs in liquids and gases.
Method of heat transfer transmitted through empty space.

Term
chemical
conservation
energy transfer
energy transformation
nuclear
conduction
heat
Joule
kinetic
convection
rating
light
sound
elastic potential energy
gravitational potential energy
electrical
efficiency
radiation

RATE MY UNDERSTANDING
Shade the face that shows your rating  

5.10 Thinking about my learning

1 Tick the square that best matches your understanding for each of the big ideas.

	Big Ideas	I still need help with this	I understand this	I understand this well and can teach someone about this
Knowing	I can identify the different forms of energy.			
	I can show how one form of energy changes to one or more different types of energy.			
	I know that energy is never lost or made.			
	I know that improving energy efficiency can reduce energy consumption.			
	I know that heat energy can be transferred by conduction, convection or radiation.			
Understanding	I could explain to someone what an energy efficiency rating is.			
	I can explain what is meant by wasted energy.			
	I can explain how sound passes from one place to another.			
	I can explain that energy can be transferred or transformed.			
	I can identify types of energy transformations using an energy flow diagram.			
Applying	I can classify key forms of energy in situations around me.			
	I can draw an energy flow diagram to show energy transformations.			
	I can calculate the energy efficiency of a device given its useful energy output and its total energy input.			

(2) Take one point from the right-hand column of the table above and write three or four sentences that you could use to explain the idea to someone else.

____

____

____

____

CHAPTER 6

Substances

6.1 Knowledge preview

Science understanding

FOUNDATION	STANDARD	ADVANCED

1 Iron is a metal that is used to build bridges and other structures. You use things made of iron every day. List three things about iron which tell you that iron is a metal.

____

____

(2) Diagrams A to E show different substances. Some are elements and some are compounds. They are all made of atoms. Atoms come in different sizes. Atoms of the same element are the same size.

A

B
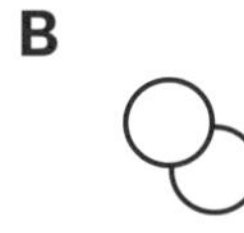
C
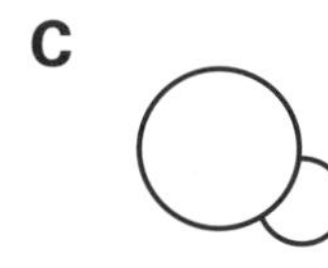
D
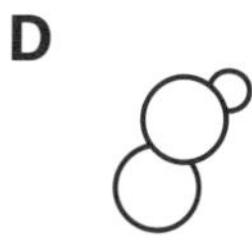
E

(a) What is an element?

____

____

(b) What is a compound?

____

____

(c) Identify the three different atoms in the diagrams A to E. Use three different colours—one for each type of atom—and colour in all of the atoms.

(d) Which atom diagrams fit the following terms? Write the atom diagram letters (A, B, C, D or E) in the spaces provided. There may be more than one diagram that fits each description.

(i) an element ____________________ **(iii)** a molecule ____________________

(ii) a compound ____________________ **(iv)** an atom ____________________

3 Scientists sometimes use prefixes to describe the number of atoms. They may use a prefix when describing substances made of more than one atom. Match the prefix with the number it represents. Complete the table using numbers 1 to 6.

Prefix	pent	di	hex	mono	tetra	tri
Number of atoms						

prefix (*n*) an affix that is added to the front of a word which changes its meaning, e.g. happy, unhappy

6.2 Periodic table quiz

Science understanding

FOUNDATION	STANDARD	ADVANCED

Scientists organise the elements in order of increasing atomic number on a grid called the periodic table (Figure 6.2.1). The periodic table helps scientists to understand some of the physical and chemical properties of elements.

H hydrogen 1																	**He** helium 2
Li lithium 3	**Be** beryllium 4											**B** boron 5	**C** carbon 6	**N** nitrogen 7	**O** oxygen 8	**F** fluorine 9	**Ne** neon 10
Na sodium 11	**Mg** magnesium 12											**Al** aluminium 13	**Si** silicon 14	**P** phosphorus 15	**S** sulfur 16	**Cl** chlorine 17	**Ar** argon 18
K potassium 19	**Ca** calcium 20	**Sc** scandium 21	**Ti** titanium 22	**V** vanadium 23	**Cr** chromium 24	**Mn** manganese 25	**Fe** iron 26	**Co** cobalt 27	**Ni** nickel 28	**Cu** copper 29	**Zn** zinc 30	**Ga** gallium 31	**Ge** germanium 32	**As** arsenic 33	**Se** selenium 34	**Br** bromine 35	**Kr** krypton 36
Rb rubidium 37	**Sr** strontium 38	**Y** yttrium 39	**Zr** zirconium 40	**Nb** niobium 41	**Mo** molybdenum 42	**Tc** technetium 43	**Ru** ruthenium 44	**Rh** rhodium 45	**Pd** palladium 46	**Ag** silver 47	**Cd** cadmium 48	**In** indium 49	**Sn** tin 50	**Sb** antimony 51	**Te** tellurium 52	**I** iodine 53	**Xe** xenon 54
Cs caesium 55	**Ba** barium 56	**La** lanthanum 57	**Hf** hafnium 72	**Ta** tantalum 73	**W** tungsten 74	**Re** rhenium 75	**Os** osmium 76	**Ir** iridium 77	**Pt** platinum 78	**Au** gold 79	**Hg** mercury 80	**Tl** thallium 81	**Pb** lead 82	**Bi** bismuth 83	**Po** polonium 84	**At** astatine 85	**Rn** radon 86
Fr francium 87	**Ra** radium 88	**Ac** actinium 89	**Rf** rutherfordium 104	**Db** dubnium 105	**Sg** seaborgium 106	**Bh** bohrium 107	**Hs** hassium 108	**Mt** meitnerium 109	**Ds** darmstadtium 110	**Rg** roentgenium 111	**Cn** copernicium 112	**Uut** ununtrium 113	**Uuq** ununquadium 114	**Uup** ununpentium 115	**Uuh** ununhexium 116	**Uus** ununseptium 117	**Uuo** ununoctium 118

Lanthanoids	**Ce** cerium 58	**Pr** praseodymium 59	**Nd** neodymium 60	**Pm** promethium 61	**Sm** samarium 62	**Eu** europium 63	**Gd** gadolinium 64	**Tb** terbium 65	**Dy** dysprosium 66	**Ho** holmium 67	**Er** erbium 68	**Tm** thulium 69	**Yb** ytterbium 70	**Lu** lutetium 71
Actinoids	**Th** thorium 90	**Pa** protactinium 91	**U** uranium 92	**Np** neptunium 93	**Pu** plutonium 94	**Am** americium 95	**Cm** curium 96	**Bk** berkelium 97	**Cf** californium 98	**Es** einsteinium 99	**Fm** fermium 100	**Md** mendelevium 101	**No** nobelium 102	**Lr** lawrencium 103

Figure 6.2.1 The periodic table

Key: H — symbol; hydrogen — name; 1 — atomic number

Use the periodic table to answer the following questions.

1 State the total number of elements listed on the periodic table. ______________________

2 Identify the chemical symbol of the following elements.

hydrogen ______________________ helium ______________________

carbon ______________________ oxygen ______________________

nitrogen ______________________ aluminium ______________________

calcium ______________________ iron ______________________

3 Name the elements with the following chemical symbols.

Li ______________________ B ______________________

Na ______________________ Si ______________________

P ______________________ Cl ______________________

Cr ______________________ Cu ______________________

4 List the names and symbols of all the elements whose names start with the letter 'C'.

5 Identify three elements named after famous scientists.

6 Identify three elements named after a place, country, continent or planet.

7 Some chemical symbols do not appear to correspond to their chemical names in English. For example, the chemical symbol for silver is Ag. List the name and symbol of five other elements whose chemical symbols do not correspond with the names of the elements.

(8) In the table below, list five elements that you might use in your everyday life and identify where they might be used.

Element	Uses

RATE MY UNDERSTANDING
Shade the face that shows your rating  

6.3 Element crossword

Science understanding

FOUNDATION | STANDARD | ADVANCED

1 Refer to the periodic table on page 80 to complete the crossword below by filling in the element name that corresponds to each symbol.

Across		Down	
8	Al	**1**	Pt
9	Ti	**2**	Be
11	O	**3**	Li
13	B	**4**	Cl
15	Cu	**5**	Ar
17	N	**6**	Na
18	P	**7**	C
22	Ca	**10**	Mg
23	F	**12**	K
24	Fe	**14**	Si
26	He	**16**	Au
		19	S
		20	H
		21	Ag
		25	Ne

RATE MY UNDERSTANDING
Shade the face that shows your rating

6.4 Identifying elements

Science understanding

FOUNDATION | **STANDARD** | ADVANCED

1 Match the elements below to the properties listed in the table. Use what you know about the elements in the box to help you do the matching.

aluminium Al	carbon C	chlorine Cl	copper Cu	gold Au
helium He	iron Fe	nitrogen N	oxygen O	sulfur S

Elements and their properties

Description of properties	Chemical name	Chemical symbol
(a) I am lightweight and shiny and conduct electricity very well. For these reasons, I am used in overhead power lines. I am also used in soft-drink cans because I can be recycled.		
(b) At room temperature I am a solid, bright yellow powder. I am a typical non-metal. I don't conduct electricity and I crumble easily. I can be found under oxygen on the periodic table.		
(c) I can be found in many different forms. Sometimes I am a black crumbly solid called charcoal. However, I can also form very hard, beautiful and expensive crystal lattices called diamond.		
(d) I am a colourless, odourless gas that makes up most of the air you breathe but I am not oxygen. I am one of the first 10 elements listed in the periodic table.		
(e) I am a yellow gas with a pungent smell. Don't breathe me in or I will damage your lungs. I am also used in swimming pools to kill bacteria. I am between elements 10 and 20 on the periodic table.		
(f) I am yellow and shiny. I conduct electricity very well so am sometimes used for wiring in electrical equipment. However, I am more commonly used in jewellery because I am rare and expensive.		
(g) I am strong and hard and can be bent into many different shapes. That is why I am used in construction. However, I am often mixed with metals and carbon to prvent me rusting.		
(h) I am a very light and non-toxic gas. I do not react with other substances so I am often used to make party balloons that float.		
(i) I am an invisible, non-toxic gas. I am one of the most important elements on Earth. I am in water, sand and air. You need me to breathe and stay alive. Plants produce me through photosynthesis.		
(j) I am shiny and orange-brown in colour. I can be drawn into wires or hammered into sheets. I conduct electricity very well, which makes me perfect for household wiring and electrical equipment.		

RATE MY UNDERSTANDING
Shade the face that shows your rating  

6.5 Ozone and the ozone hole

Science inquiry skills

FOUNDATION	STANDARD	ADVANCED

Processing & Analysing | Questioning & Predicting | 

The element oxygen (O) can have two different forms. The most common is oxygen gas (O_2), which is made up of molecules with two oxygen atoms. However, oxygen atoms can also form molecules with three oxygen atoms. This form of oxygen is known as ozone (O_3). The different arrangement of atoms for oxygen gas and ozone are shown below.

stratosphere (*n*) the level of the Earth's atmosphere extending to 50 km above the Earth's surface

ultraviolet light (*n*) rays of light that cannot be seen but can cause burns to skin like sunburn

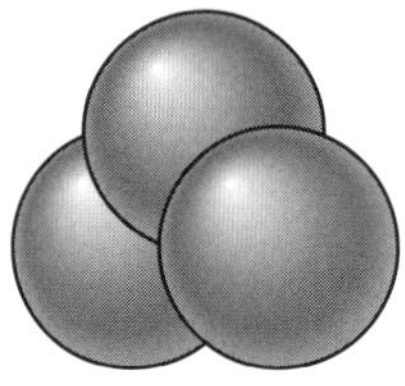

Ozone O_3

Oxygen gas and ozone have very different properties. Oxygen is colourless, odourless and non-toxic whereas ozone is a pale blue gas with a strong smell that irritates your eyes and throat. Ozone also absorbs ultraviolet radiation.

Most ozone is found in the stratosphere, about 10 to 50 km above the Earth's surface in a layer known as the ozone layer. The ozone layer absorbs some of the ultraviolet light from the Sun, protecting us from these rays.

In the mid-1980s it was discovered that pollution was making the ozone layer very thin near the North and South Poles resulting in an ozone hole. The Earth's natural protection was being destroyed, which could increase the likelihood of people developing skin cancer.

Table 6.5.1 shows the amount of ozone above the South Pole as recorded by NASA every two years. The amount of ozone is measured in Dobson Units (DU). This unit of measurement was developed specifically to measure ozone in the atmosphere.

Table 6.5.1 Ozone in South Pole atmosphere 1980–2014

South Pole ozone levels			
Year	**Ozone minimum (DU)**	**Year**	**Ozone minimum (DU)**
1980	203	1998	99
1982	185	2000	98
1984	164	2002	157
1986	158	2004	124
1988	171	2006	98
1990	124	2008	114
1992	114	2010	128
1994	92	2012	139
1996	108	2014	129

6.5 Ozone and the ozone hole

(1) Construct a line graph using the axes provided to show how the level of ozone has varied from 1980 to 2014. Write a heading for the graph and number the scale along the vertical axis (y).

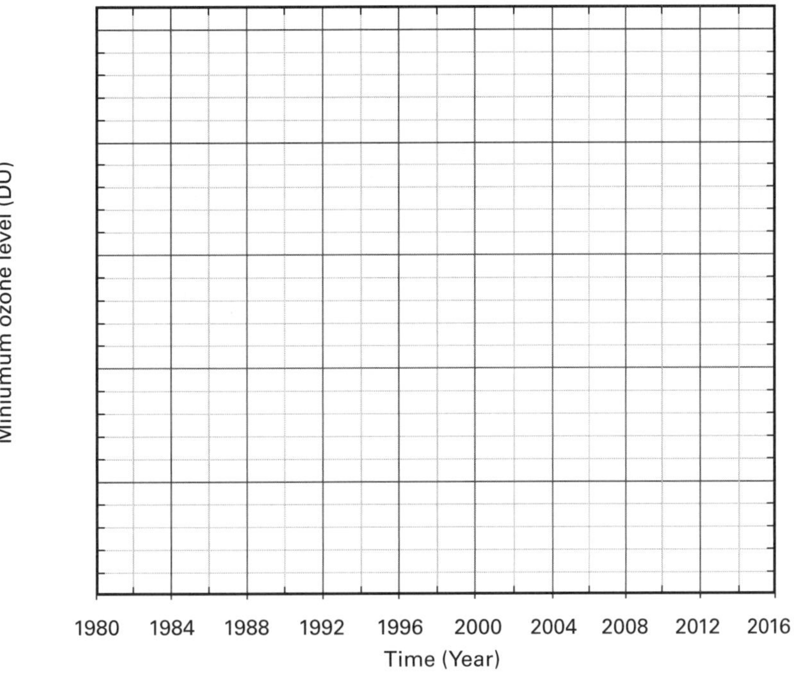

(2) Identify from the graph:

(a) the minimum ozone level and the year in which it occurred ____________

(b) the years where there were increases in ozone levels compared with the previous two-year reading

____

(c) the four years when people would be most at risk from ultraviolet radiation

____

(d) the ozone level in 1989 ____________

(e) the number of years with ozone readings over 100 DU

____

(f) the 4 years showing the lowest health risk due to ultraviolet radiation

____

(g) the year in which the ozone was recorded at 139 DU.

____

3 Predict what the minimum level of ozone might be this year based on the data on the graph. Justify your answer.

____

____

RATE MY UNDERSTANDING
Shade the face that shows your rating  

6.6 Molecules and lattices

Science understanding

FOUNDATION | STANDARD | **ADVANCED**

The elements and compounds found in the world around you can exist as single atoms, molecules or large grid-like structures called crystal lattices. Sometimes all the atoms in these molecules and lattices are the same element. Sometimes the atoms in molecules and lattices are different elements grouped together to form compounds.

grid (*n*) a framework of squares

lattice (*n*) an arrangement in space of isolated points in a regular pattern showing positions of atoms or molecules in a crystal structure

Molecules

Molecules are groups of atoms joined together. The molecules that make up pure substances such as elements and compounds are all identical.

The molecular formulas for elements and compounds tell you which type of atoms are in the molecule and how many of each type there are. For example, the molecular formula for the element oxygen is O_2, which means that each molecule contains two oxygen atoms. The molecular formula for the compound carbon dioxide is CO_2, which means that there is one carbon atom and two oxygen atoms in each carbon dioxide molecule.

(1) Construct diagrams of the following molecules.

Oxygen O_2	Carbon dioxide CO_2	Nitrogen N_2
Water H_2O	Ozone O_3	Carbon monoxide CO
Phosphorus P_4	Methane CH_4	Hydrogen peroxide H_2O_2

Lattices

Crystal lattices such as diamond or sodium chloride are made up of a huge number of atoms stuck together in large grid-like structures. For this reason, crystal lattices do not have molecular formulas. Instead they are referred to by their chemical formulas. The chemical formula of a lattice tells you which type of atoms make up the lattice and the ratio of each type of atom in the lattice. For example, the chemical formula for sodium chloride (table salt) is NaCl. This means that in the crystal lattice there is one sodium atom for every chlorine atom. For silicon dioxide (beach sand), the chemical formula of SiO_2 means that for every silicon atom in the lattice there are two oxygen atoms.

The crystal lattices of elements are made up of only one type of atom so their chemical formulas are exactly the same as the chemical symbols for the elements. For example, diamond is a crystal lattice made up of only carbon atoms, so its chemical formula is just C.

(2) Construct diagrams to represent the chemical formulas of gold and magnesium oxide.

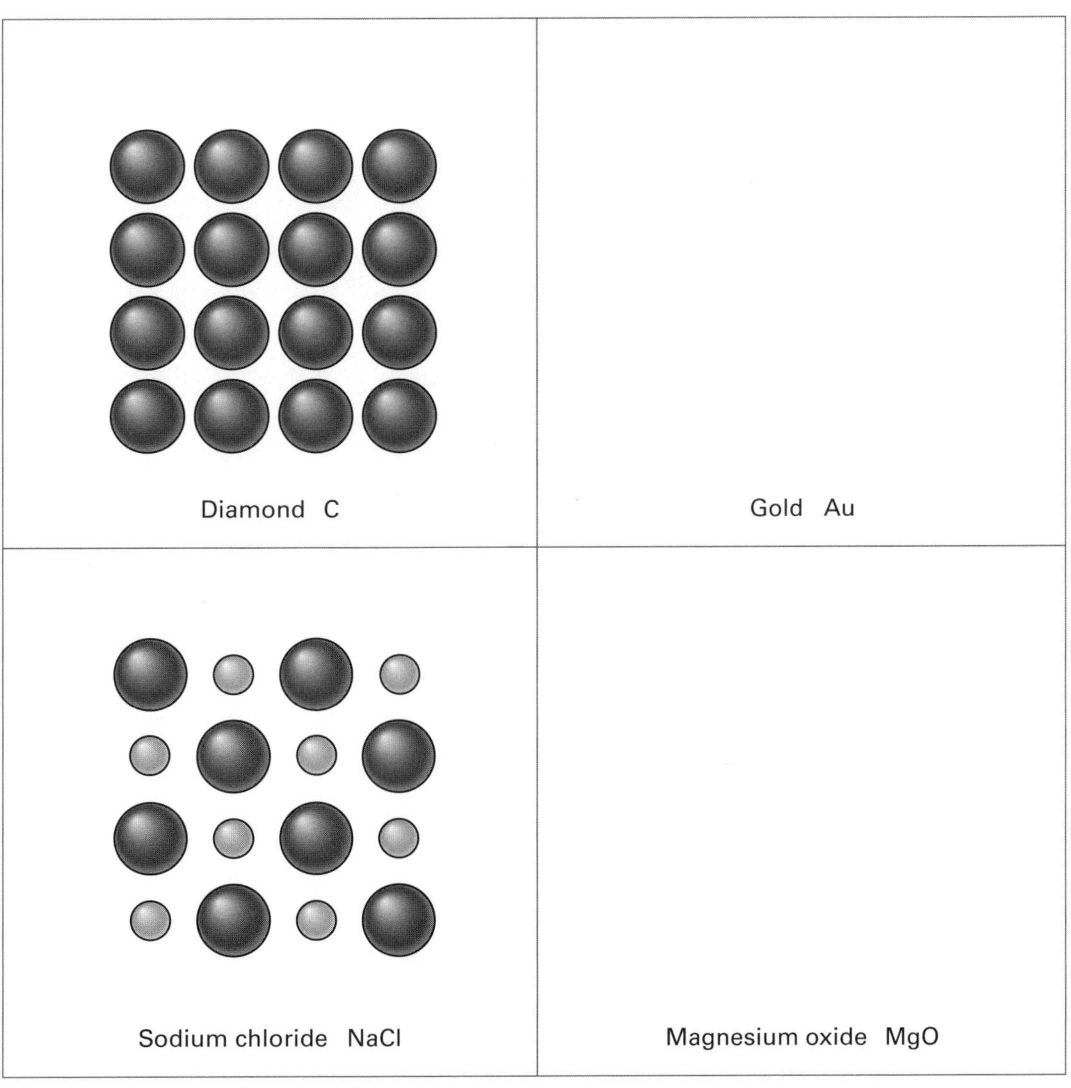

6.7 Air mixture

Science understanding

FOUNDATION	**STANDARD**	ADVANCED

The air you breathe is a mixture of elements and compounds. It contains approximately 78% nitrogen (N_2), 21% oxygen (O_2) and 1% argon (Ar). It also contains very small amounts of carbon dioxide (CO_2), neon (Ne), helium (He) and methane (CH_4). The box below contains 78 molecules of nitrogen (N).

1. Identify how many molecules of oxygen and argon are required to make this a box of air. Add the molecules of oxygen and argon to the diagram using red for oxygen and blue for argon.

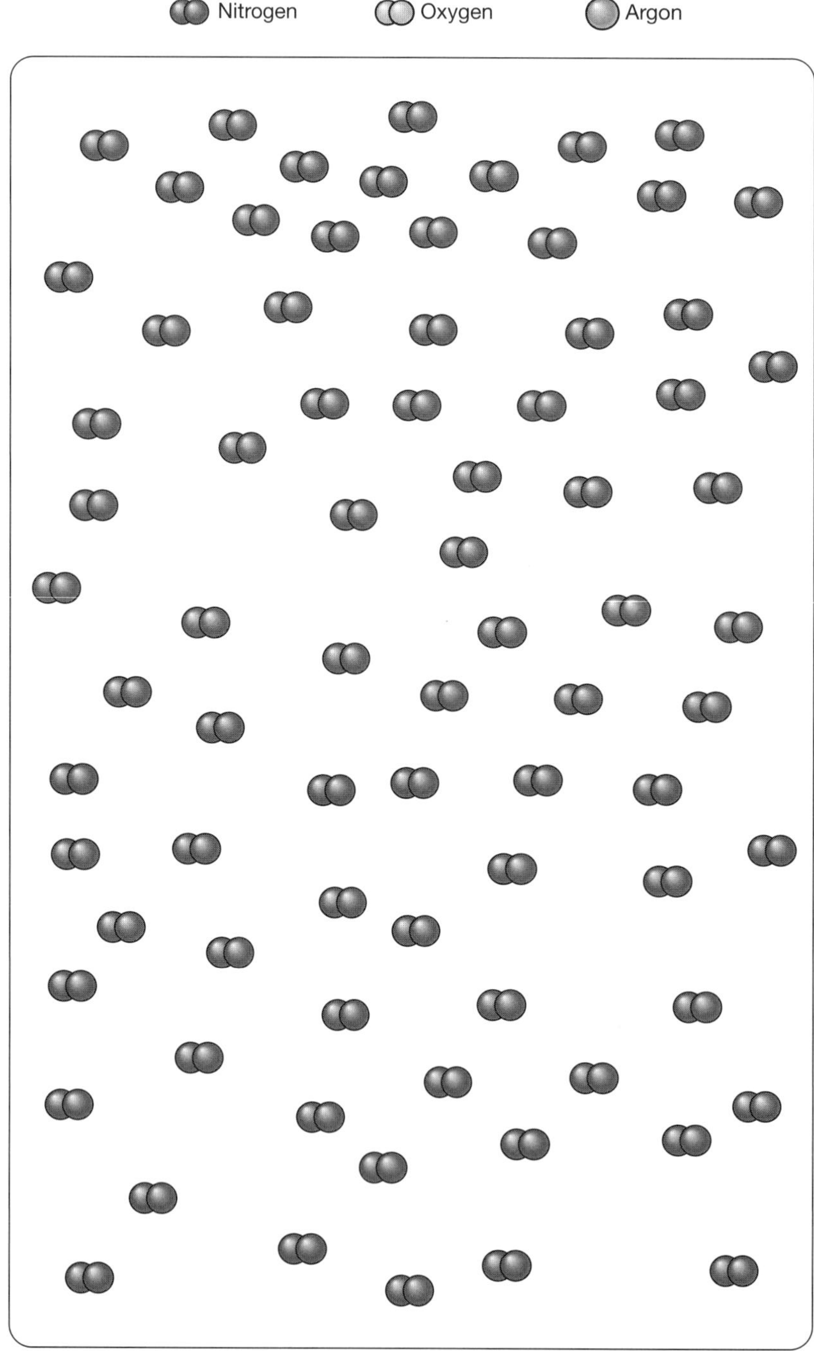

RATE MY UNDERSTANDING
Shade the face that shows your rating

6.8 Finding mine sites

Science inquiry skills

FOUNDATION | STANDARD | **ADVANCED**

Processing & Analysing | Communicating | Evaluating

Geochemists searching for new nickel deposits decided to take soil samples from the area shown in the map. This map shows the magnetic anomalies in the area. There is a stream flowing through the site (Figure 6.8.1).

anomaly (*n*) something odd or does not fit in, an abnormality
stream (*n*) little river

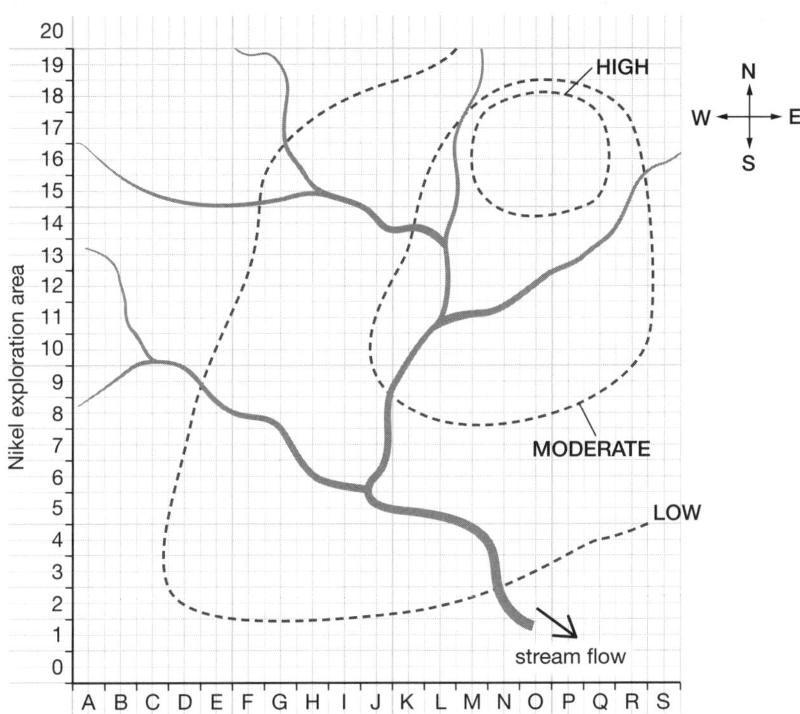

Figure 6.8.1 Map showing magnetic anomalies (from high to low magnetism) in an area being explored

1. Describe the method probably used by geophysicists that helped them decide that this area was worth sampling.

____

____

2. Explain why the geochemists would not have obtained useful information by sampling the water in the stream to detect minerals in the area.

____

____

3. Where would the geochemists have decided to take soil samples from, in this area? Explain why you chose that area.

____

____

4 After studying the map, the geochemists took their soil samples as shown by the grid references on the table below.

(a) Identify the part of the map where soil samples were taken and rule a black line around this area on the map.

(b) The samples returned the results for nickel shown in the table below. The values are parts per million (ppm). Anything less than 30 ppm was recorded as zero.

Quality of nickel from soil samples

Grid reference	8	9	10	11	12	13	14	15	16	17
M	0	0	30	0	0	0	0	0	0	0
N	0	35	37	59	71	74	78	109	77	35
O	0	35	37	78	87	100	140	150	135	80
P	0	35	61	66	69	85	137	100	76	55
Q	0	0	0	31	33	30	31	30	0	0

Classify and colour these values for nickel grade as low (red) if they are in the range 30–50 ppm, moderate (yellow) if they are 51–100 ppm or high (green) if they are over 101 ppm. Leave zeros blank. Do this by colouring the above table.

5 Modify the map shown in Figure 6.8.1 to show the results of nickel grade from the soil samples. Do this by shading the map, using the same colours used for question 4b. Add a legend to the side of the map to explain the colouring. (Note that the H, M and L areas on the map indicate areas of high, moderate and low magnetism—not grades of nickel.)

6 What do you think geochemists would do next and why do you think they would do this?

7 Propose the best location to begin mining for nickel. Justify your choice by explaining why this is the best location and why other locations are less suitable.

RATE MY UNDERSTANDING
Shade the face that shows your rating

6.9 Literacy review

Science understanding

FOUNDATION | **STANDARD** | ADVANCED

Recall your knowledge of substances by choosing words from the box to complete the statements below. Some words may be used more than once. Some words may not be used at all. Two have been completed for you.

monatomic (*adj*) made up of single atoms

atoms ✓	bauxite	break	compounds	conduct	diamond
elements	gas	hardness	lattices	liquid	mixtures
molecules ✓	nucleus	ores	solid	talc	

1. ___Atoms___ are the smallest building blocks that make up all the substances around you. Substances made up of just one type of atom are known as ______________ .
2. Metallic elements are shiny, ______________ electricity and heat, and can be drawn into wires or hammered into sheets. They are usually ______________ at room temperature.
3. Non-metallic elements are usually dull, do not conduct electricity or heat and ______________ when a force is applied. Most non-metals are solid or ______________ at room temperature.
4. The atoms that make up the elements can be monatomic, in clusters called ___molecules___ or in large crystal ______________ .
5. Pure substances made up of more than one type of atom are known as ______________ . These substances can be made up of atoms in crystal lattices.
6. Substances that are made up of a combination of different elements and compounds are known as ______________ .
7. Minerals can be identified by their colour, lustre and ______________ .
8. Rocks that contain valuable minerals such as ______________ are called ______________ .
9. Mohs hardness scale lists ______________ as the softest mineral and ______________ as the hardest mineral.

RATE MY UNDERSTANDING
Shade the face that shows your rating   

6.10 Thinking about my learning

1 In this box, write down four ideas from this chapter on substances that you feel you understand well.

2 In this box write down four ideas about substances that you feel you don't understand as well as you would like to.

3 **(a)** Find someone in your class who needs help with one of the ideas which you understand well and explain the idea to them.

(b) Find a member of your class who understands one of the ideas which you still don't feel confident about and ask them to explain the concept to you.

CHAPTER 7

Physical and chemical change

7.1 Knowledge preview

Science understanding

FOUNDATION | STANDARD | ADVANCED

1 Look carefully at the diagram below.

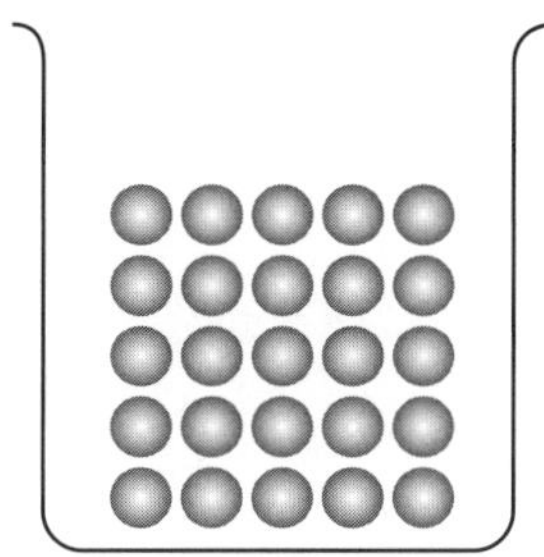

(a) Use the empty beaker to re-draw the particles of the melted solid after heat has been applied to the beaker.

(b) Is the change that you showed a physical change or a chemical change? Explain your answer.

---

2 Identify each of the following as a physical change (P) or a chemical change (C) by writing P or C in the space provided.

(a) Ice melts to water. ________

(b) Glucose breaks down to carbon dioxide and water during respiration. ________

(c) An aluminium can is squashed to take up less space in the rubbish bin. ________

(d) Sodium bicarbonate releases carbon dioxide to make the cake rise. ________

(e) An egg is cooked. ________

3 Complete the flow chart using the terms: melting, freezing, condensation, evaporation. The first one has been done for you.

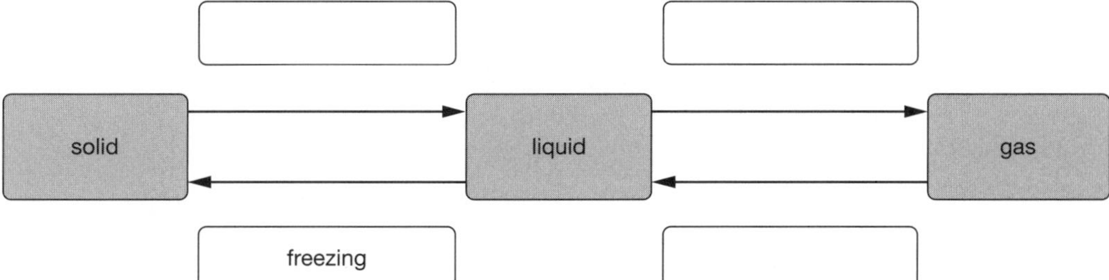

7.2 Physical and chemical change

Science understanding

FOUNDATION	STANDARD	ADVANCED

The world around you is constantly changing. Some of these changes are known as physical changes, in which there are no new substances produced. Other changes are known as chemical changes, in which new substances are produced.

1 Classify each of the following situations as either a physical or chemical change by ticking the correct box.

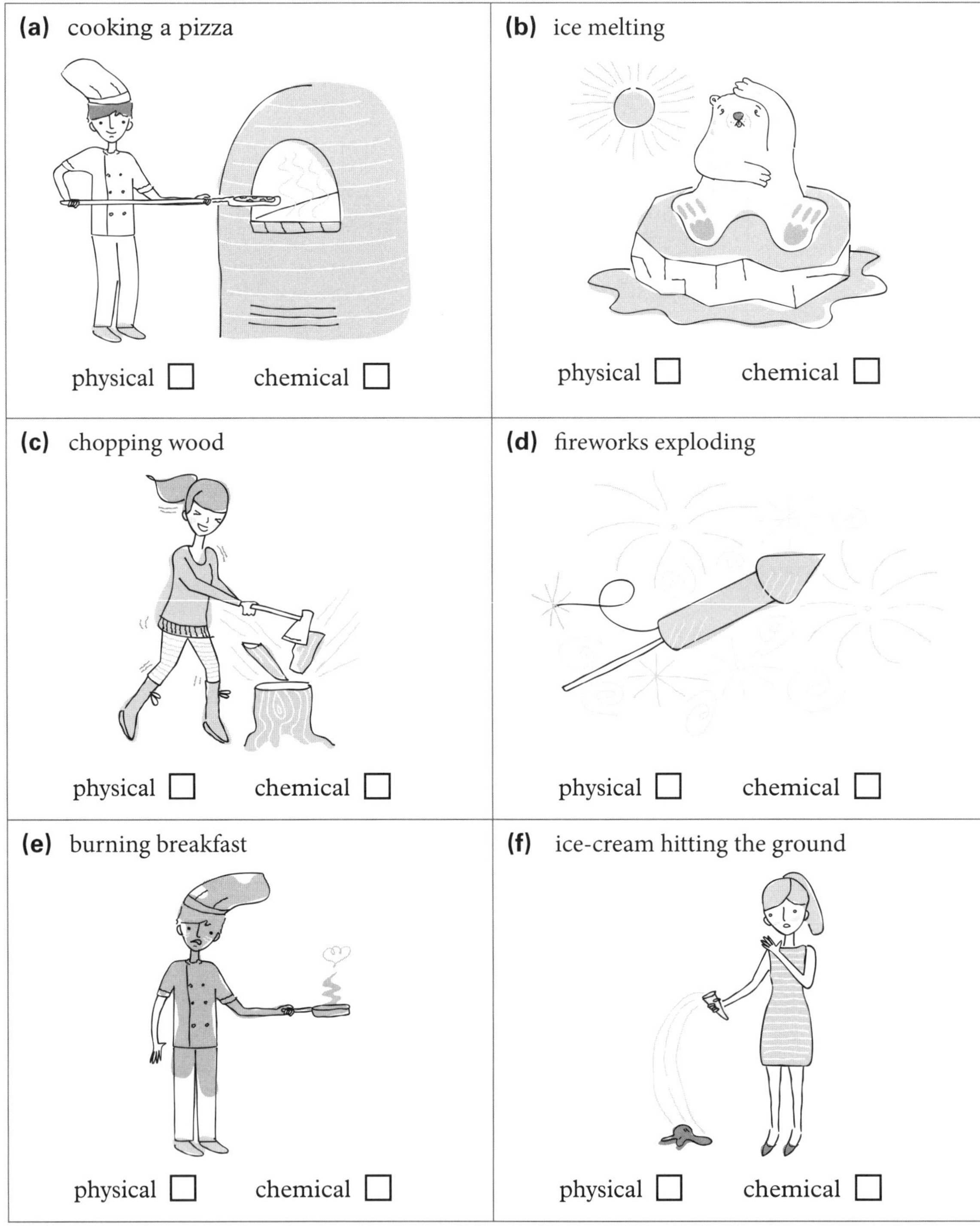

(a) cooking a pizza physical ☐ chemical ☐	**(b)** ice melting physical ☐ chemical ☐
(c) chopping wood physical ☐ chemical ☐	**(d)** fireworks exploding physical ☐ chemical ☐
(e) burning breakfast physical ☐ chemical ☐	**(f)** ice-cream hitting the ground physical ☐ chemical ☐

(g) stirring a cake mixture

physical ☐ chemical ☐

(h) crashing a car

physical ☐ chemical ☐

(i) autumn leaves changing colour

physical ☐ chemical ☐

(j) an explosion

physical ☐ chemical ☐

(k) making a new chemical

physical ☐ chemical ☐

(l) baking bread

physical ☐ chemical ☐

(m) boiling water

physical ☐ chemical ☐

(n) breaking a plate

physical ☐ chemical ☐

RATE MY UNDERSTANDING
Shade the face that shows your rating
☺ 😐 ☹

7.3 Recycling glass

Science understanding

FOUNDATION | STANDARD | ADVANCED

Glass makes up 7% of domestic waste. Fortunately, glass can be recycled over and over again. During the recycling process, glass goes through several physical changes. The stages listed below are jumbled.

- Each type of glass is crushed into small pieces.
- The small pieces of glass are melted in a furnace to make liquid glass, which is then poured into moulds.
- The glass is allowed to solidify into its new form, such as bottles, plates or windows.
- The glass is thrown into the bin and it gets mixed in with the rest of the household rubbish.
- The glass is separated from the general rubbish.
- The glass is sorted by colour.

1 Complete the flow chart below. Refer to the jumbled list of the glass recycling process above. Arrange these processes in the correct order on the flow chart. Begin at the centre top of the flow chart with the household glass waste thrown into a bin. Draw a diagram to represent each stage in the process.

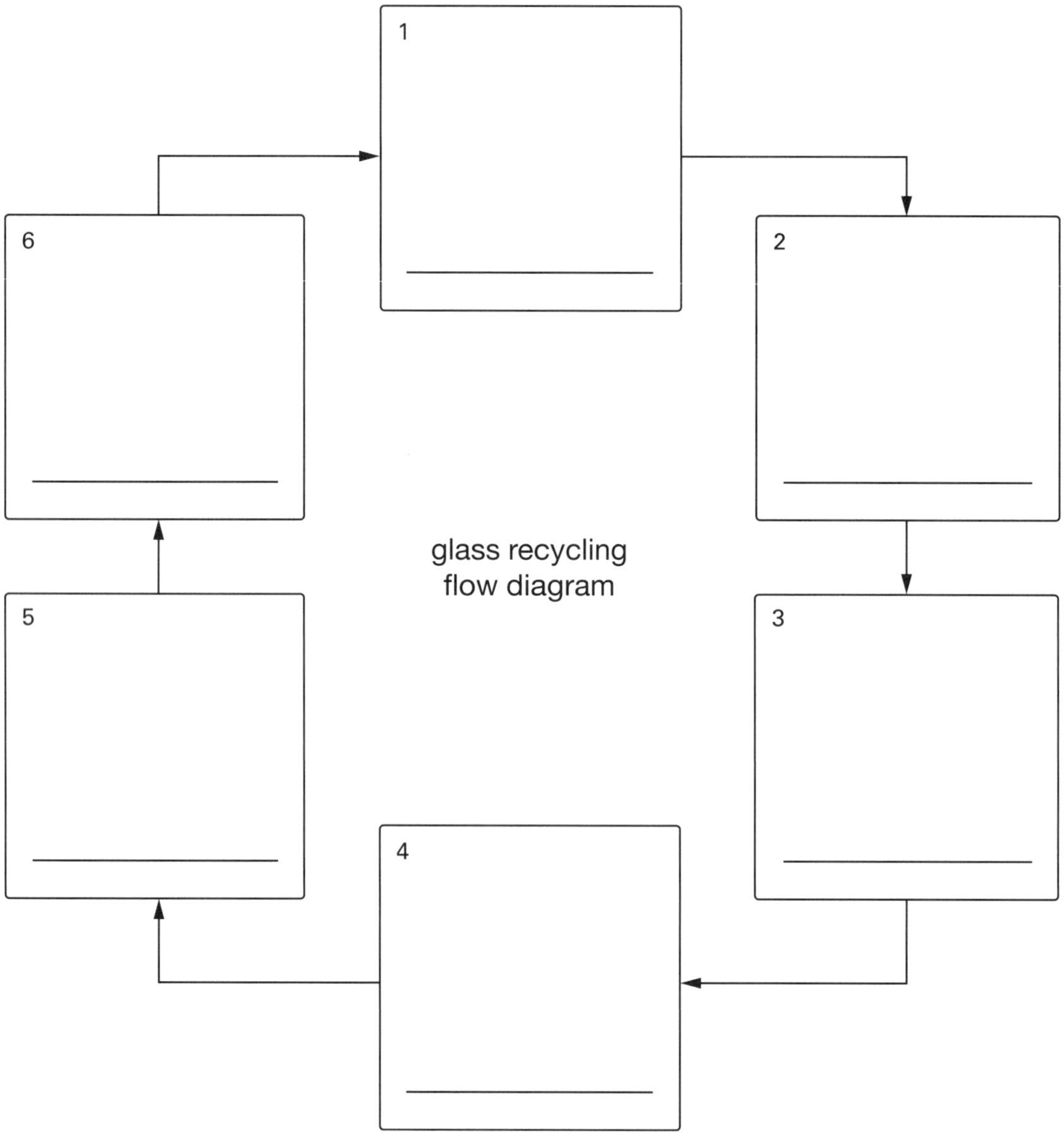

1 ______________

2 ______________

3 ______________

4 ______________

5 ______________

6 ______________

glass recycling flow diagram

RATE MY UNDERSTANDING
Shade the face that shows your rating

7.4 Boyle's law

Science as a human endeavour

FOUNDATION	STANDARD	ADVANCED

Robert Boyle (Figure 7.4.1) was born in Ireland in 1627 and was considered to be an alchemist. Like all alchemists, he believed that base metals, such as lead, could be turned into gold. However, unlike other alchemists, he approached his work with strict scientific method and criticised other alchemists for their careless approach. For this reason, he can also be considered one of the world's first chemists.

Boyle was particularly interested in studying the pressure of gases. The pressure of a gas is how much force the gas puts on its container. Boyle discovered that the pressure of a gas changed when the size of its container (volume) changed (Figure 7.4.2). In one of his most famous experiments, Boyle measured the pressure of a gas as the volume of the container was decreased. He also made sure the temperature stayed the same.

Figure 7.4.1 Robert Boyle

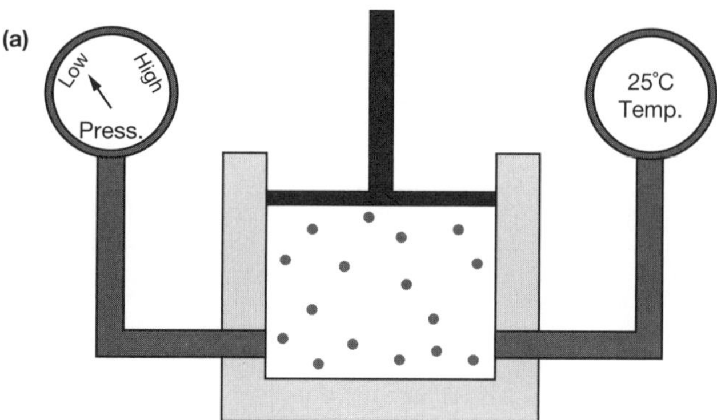

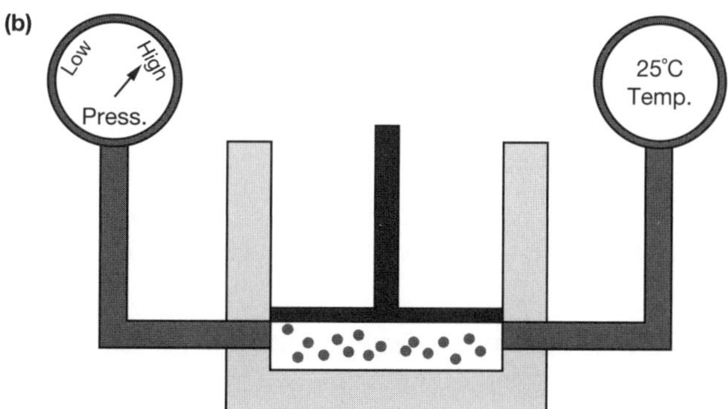

Figure 7.4.2 The effect of volume on pressure. (a) When the volume is large, the pressure is low. (b) When the volume is small, the pressure is high.

7.4 Boyle's law

Results like those that Boyle may have observed are shown in this table.

Changes in pressure as volume changes										
Volume (L)	10	9	8	7	6	5	4	3	2	1
Pressure (kPa)	100	111	125	143	167	200	250	333	500	1000

1. Use the data from the table to construct a line graph on the axes below.

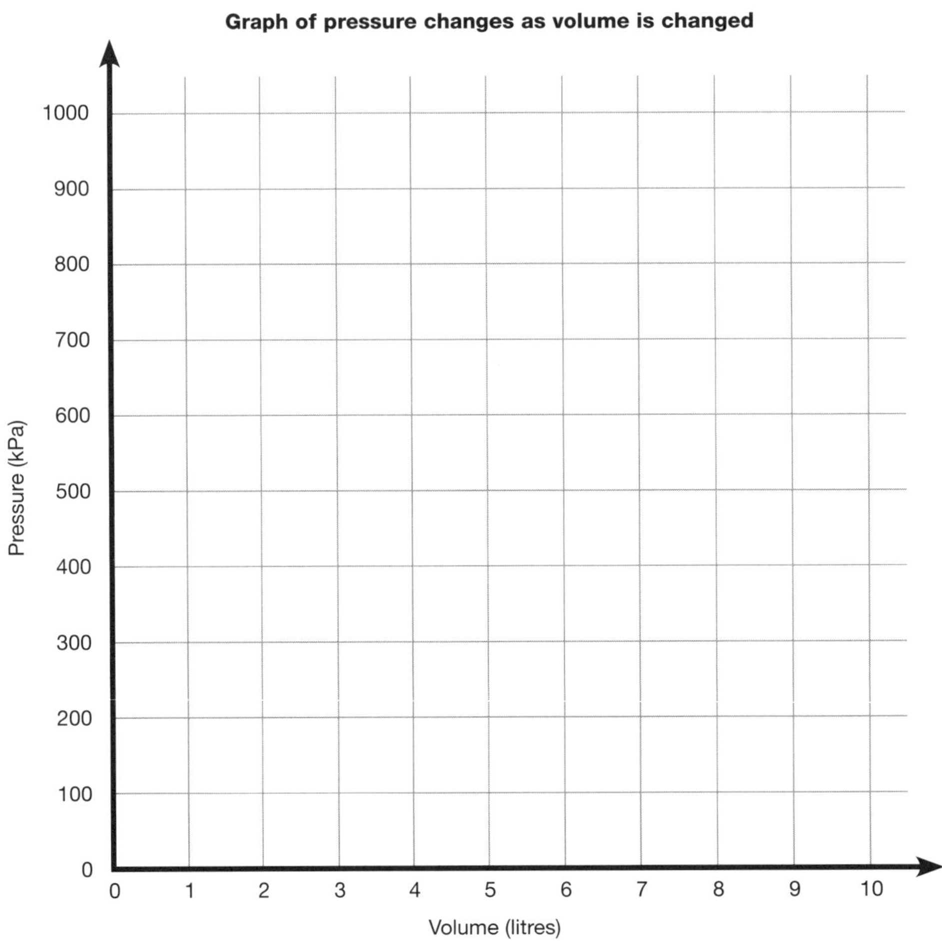

2. Circle the answer that completes each sentence.

(a) as the volume increases, the pressure: increases / decreases / stays the same

(b) as the pressure increases, the volume: increases / decreases / stays the same

(c) when the volume doubles, the pressure: doubles / halves / stays the same

(d) when the pressure doubles, the volume: doubles / halves / stays the same

RATE MY UNDERSTANDING
Shade the face that shows your rating

7.5 The particle model

Science understanding

FOUNDATION | **STANDARD** | ADVANCED

Scientists use models to help understand and predict how things work. Models are simplified versions of the real thing that explain the most important features.

A very important model in science is the particle model. The particle model is used by scientists to understand the physical properties of solids, liquids and gases. The particle model makes three assumptions:

1. Solids, liquids and gases are all made up of hard, ball-like particles that cannot be split (are indivisible) and are invisible to the naked eye.
2. These particles are constantly moving and/or vibrating.
3. There are forces of attraction between the particles.

assumption (*n*) a good guess, something you expect to be true

indivisible (*adj*) cannot be divided

invisible (*adj*) cannot be seen

naked eye (*n*) the eyes alone without a microscope

The diagrams in Figure 7.5.1 are of solids, liquids and gases as described by the particle model (not to scale).

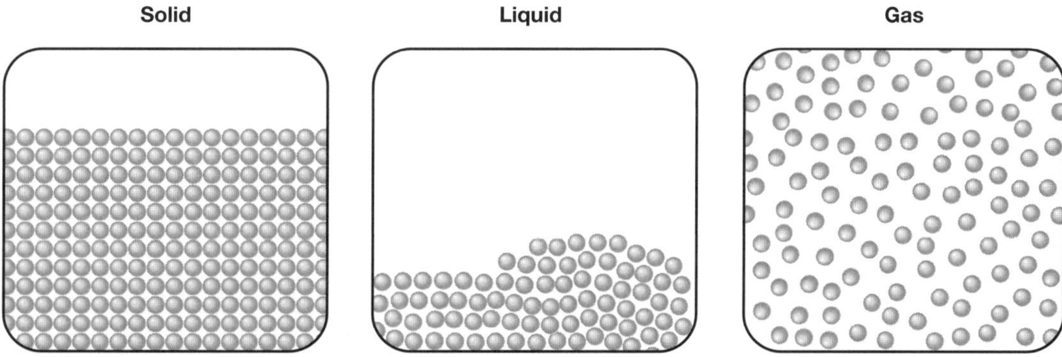

Figure 7.5.1 The particle model diagrams

1 Refer to the particle model diagrams above and compare the movement of particles in solids, liquids and gases.

____

____

____

2 **(a)** The forces of attraction hold the particles of substances together. Deduce in which state of matter (solid, liquid or gas) these forces are:

(i) strongest ________________________________

(ii) weakest. ________________________________

(b) Justify your answers.

____

____

____

7.5 The particle model

③ From your answers to questions 1 and 2, propose why solids hold their shape, liquids take on the shape of their container, and gases fill their container.

④ Use the particle model diagrams to describe the distance between particles in solids, liquids and gases.

⑤ Explain why solids and liquids are incompressible (cannot be squashed) while gases are compressible.

⑥ Use the particle model to draw particle model diagrams of the following scenarios.

(a) ice cube floating in water	**(b)** bubble of air trapped in a block of ice
(c) boiling water	**(d)** water droplet suspended in air

RATE MY UNDERSTANDING
Shade the face that shows your rating

7.6 Reverse processes

Science understanding

FOUNDATION	STANDARD	ADVANCED

1 For each physical change in the left column, identify the name of its reverse change in the right column and draw a line between the two. Refer to Figure 7.6.1 to help you.

Physical change
(a) condensing
(b) freezing
(c) separating
(d) crystallising
(e) expansion
(f) evaporating
(g) mixing
(h) contraction
(i) melting
(j) dissolving

Reverse change
freezing
expansion
separating
evaporating
dissolving
mixing
contraction
crystallising
melting
condensing

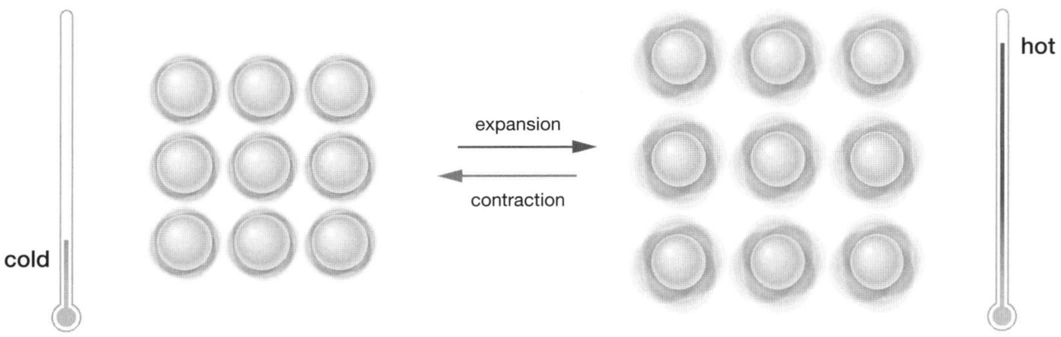

Figure 7.6.1 Expansion and contraction changes in solids and liquids

7.7 Particle diagrams of chemical and physical change

Science understanding

FOUNDATION | **STANDARD** | ADVANCED

During a chemical change, new substances are produced. For example, when carbon is heated with oxygen it produces a new substance—carbon monoxide. This is a chemical change.

During a physical change, the substances before and after the change are the same—even though they might be in a different form. For example, when liquid water boils it becomes water vapour (steam) so boiling is a physical change.

The particle model can be used to show the differences between physical and chemical changes by looking at what is happening to the particles in the substances. During a chemical change, the particles that make up the original substances rearrange to form different substances. In a physical change, the particles before and after the change are the same.

1. Determine if the particle diagram below indicates a chemical or a physical change. Justify your answer.

Before

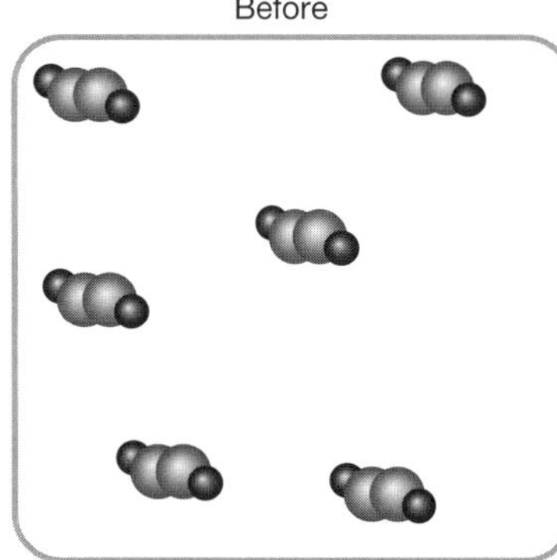

After

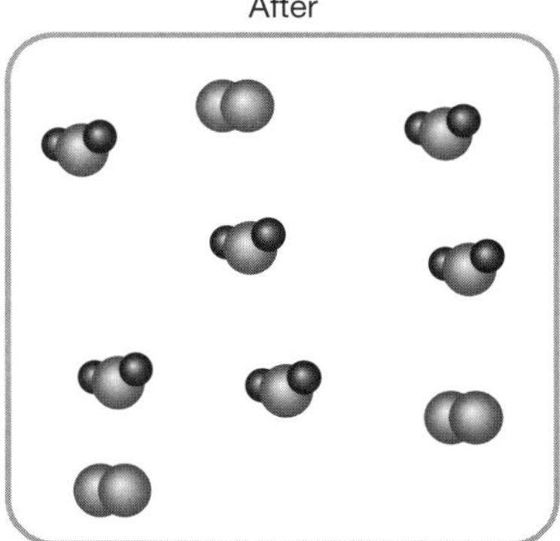

____

____

2. Determine if the particle diagram below indicates a chemical or a physical change. Justify your answer.

Before

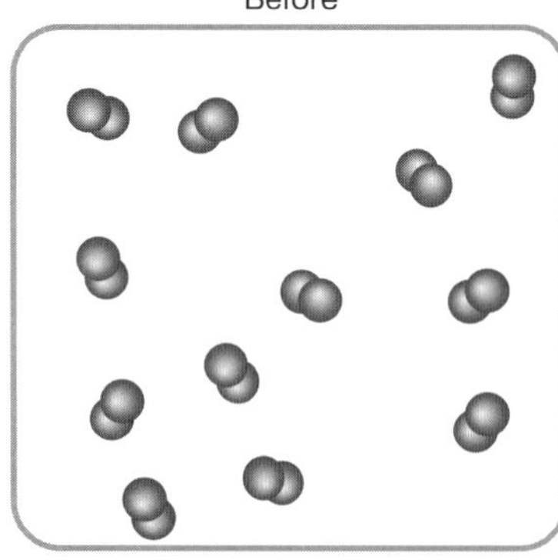

After

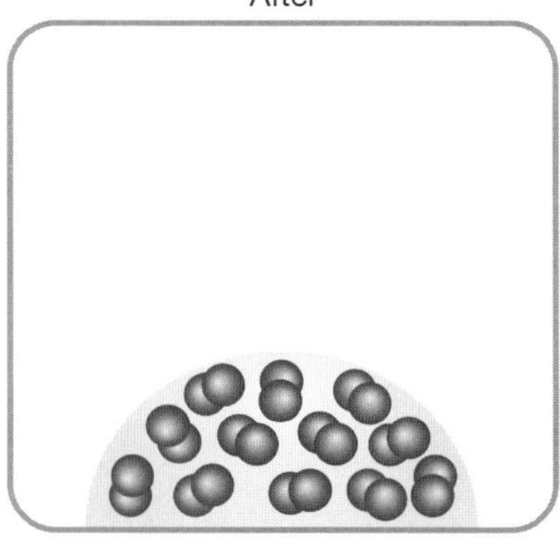

____

____

7.7 Particle diagrams of chemical and physical change

(3) When solid sulfur (S) is heated with oxygen gas (O_2), they chemically react to produce molecules of sulfur dioxide gas (SO_2). Each sulfur dioxide molecule contains one sulfur atom and two oxygen atoms. Use this information and the particle model to complete the particle diagram below.

Before the chemical reaction

After the chemical reaction

sublimation (*n*) In chemistry, sublime means to change state from a solid to a gas. It is a direct process that does not pass through a liquid state.

deposition (*n*) In chemistry, deposition is the reverse process of sublimation. It is the change of state from a gas to a solid without passing through the liquid state.

(4) When ice is heated it melts to form liquid water. Use the particle model to complete the particle diagram below to show what happens to the water particles in a block of ice after it melts.

Before melting

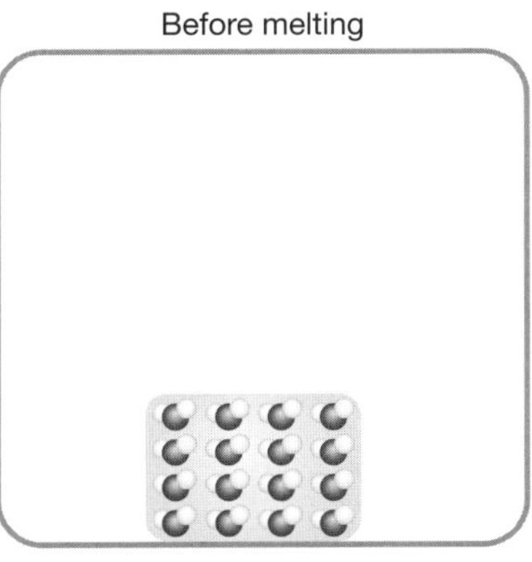

After melting

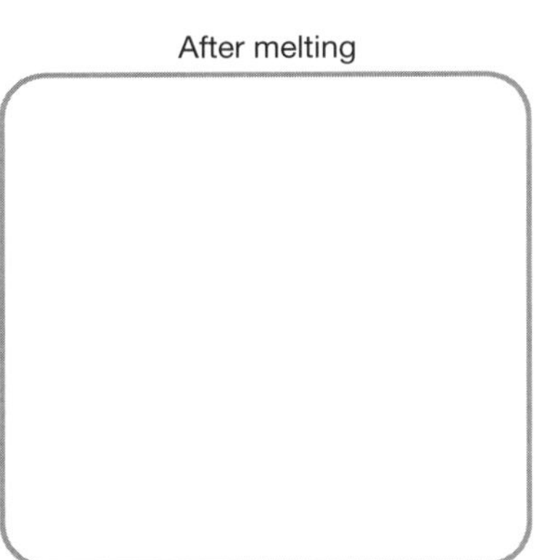

(5) Graphite is a solid that is made up of just carbon atoms. When graphite is heated to 3642°C it sublimes and the carbon atoms form a gas. Use the particle model to complete the particle diagram below to show what happens to the particles of carbon in graphite before and after it sublimes.

Before sublimation

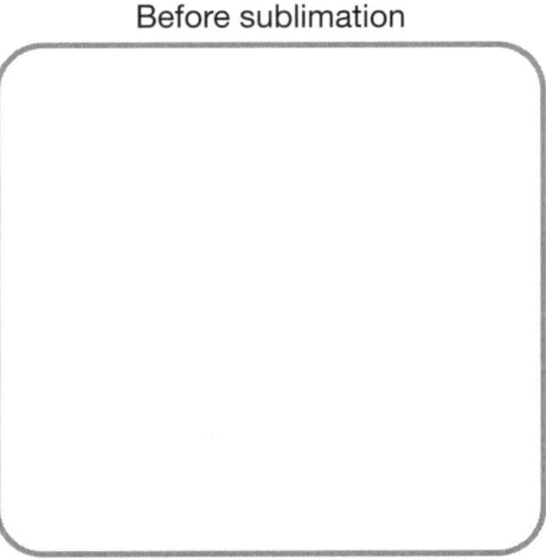

After sublimation

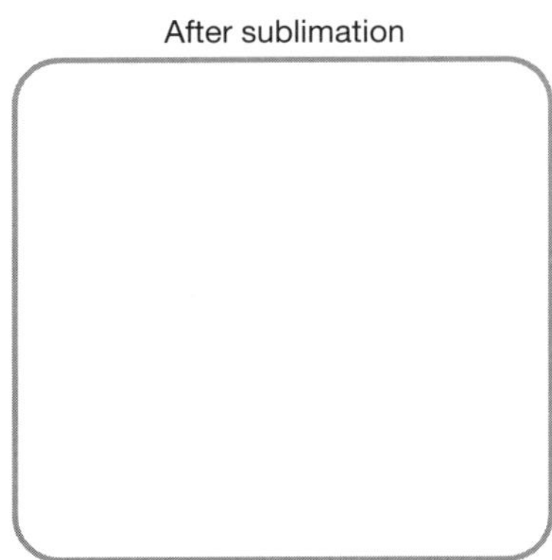

RATE MY UNDERSTANDING
Shade the face that shows your rating

7.8 Combustion for electrical power generation

Science inquiry skills

FOUNDATION	STANDARD	ADVANCED

Processing & Analysing | Questioning & Predicting | Communicating

Combustion is the term used to describe any chemical reaction where a substance burns in oxygen—usually producing large amounts of heat and light. Combustion reactions play an important role in society and your everyday life. People use combustion every day to power their vehicles, heat their homes and cook their food.

One of the most important uses of combustion is in the generation of electrical energy. Most of Australia's electrical energy comes from the combustion of fossil fuels and only a small amount comes from alternative, renewable sources. The table below lists how much different energy sources contribute to Australia's electrical energy production.

fossil fuels (*n*) energy sources made from fossils such as oil, coal and natural gas

generation (*n*) production, making

renewable (*adj*) an energy source that can be made again, or never runs out

Sources of Australia's electric energy production	
Energy source	**Percentage**
coal	76.5
oil	1
natural gas	15
hydroelectricity	5
wind	1.5
solar and other	1
total	100%

(1) Construct a pie graph to display the information in the table above. Hint: to determine the angle of each pie segment, use the formula:

$$\text{Angle} = 360° \times \frac{\text{energy source percentage}}{100}$$

Worked example:

$$\text{Angle} = 360° \times \frac{5}{100}$$

Angle = 18°

Use your protractor to mark 18 degrees on the graph.

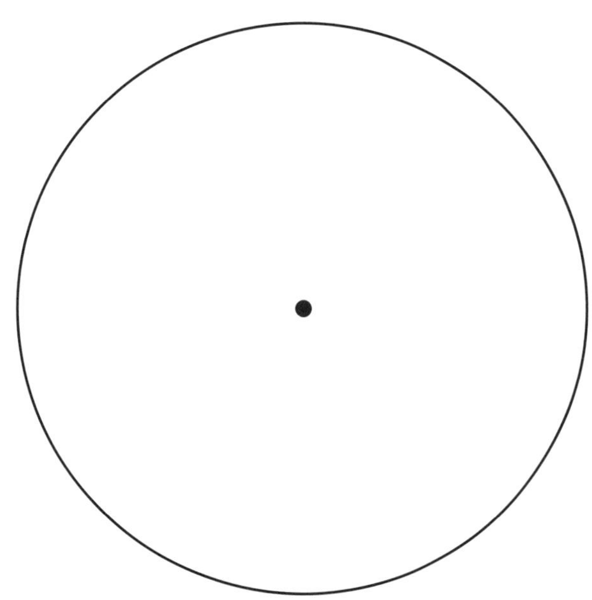

2. Calculate what percentage of Australia's electrical energy production comes from the combustion of fuels.

3. Some scientists predict that the world's fossil fuel resources may be completely used up within your lifetime. Construct a pie graph showing what you think Australia's electrical energy production might be like in 50 years.

Possible future sources of Australia's electrical energy in 50 years

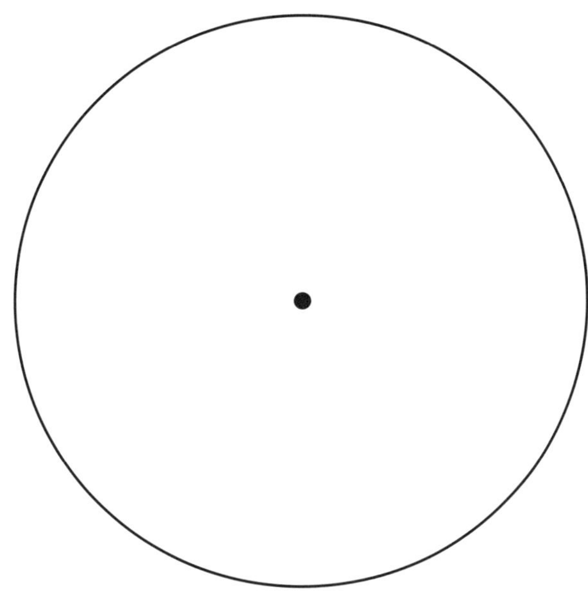

4. Justify the reasoning behind your pie graph.

RATE MY UNDERSTANDING
Shade the face that shows your rating

7.9 Physical and chemical properties

Science understanding

FOUNDATION	STANDARD	ADVANCED

Scientists use the word property to describe any feature or characteristic of a substance. Properties are usually chemical or physical. Chemical properties refer to the way in which a substance reacts chemically with other substances. Physical properties refer to all characteristics that can be measured such as colour, texture, temperature and, pressure of a substance. These physical properties can include how it looks, feels or responds to electricity, heat and light and when we apply force to it.

fishing sinkers (*n*) small, heavy metal objects on the end of a fishing line, that hold the line down
moulded (*v*) formed into a particular shape
rust (*v*) the red or orange colour that forms on old or damaged metal

1. Read through the list of descriptions below. Assess whether the description is referring to a physical or a chemical property, then place a tick in the appropriate column.

Property description	Physical property	Chemical property
(a) Aluminium is often pressed into very thin sheets to make aluminium foil.		
(b) Gold almost never reacts with acids.		
(c) Silver is a soft metal that can be moulded into different shapes to make jewellery.		
(d) Sodium metal bubbles violently when placed in water, releasing hydrogen gas.		
(e) Pure iron is not a good construction material because it rusts easily.		
(f) Lead is a very heavy metal that is used to make fishing sinkers.		
(g) Helium can be used in party balloons that float because helium gas is lighter than air.		
(h) It is important to keep methane gas away from open flames because it catches alight easily.		
(i) Chlorine gas is toxic if breathed in because it burns the nose, throat and lungs.		
(j) Diamond is the hardest natural substance, so it is used in industrial drills and saws.		
(k) Water melts at 0°C and boils at 100°C.		
(l) Copper is used extensively for electrical wiring because it conducts electricity very well.		
(m) Hydrogen peroxide (bleach) is commonly used to turn dark hair blond.		
(n) When heated, calcium carbonate will decompose into calcium oxide and carbon dioxide.		
(o) Pots and pans are made from metals because metals are very good at conducting heat.		

RATE MY UNDERSTANDING
Shade the face that shows your rating

7.10 Literacy review

Science understanding

FOUNDATION	STANDARD	ADVANCED

Recall your knowledge of physical and chemical change by choosing words from the box to complete the statements below. Cross out the words as you write them.

acidic	attracted	chemical	colour	electricity	fuels
gas	heat	moving	oxygen	particle	photosynthesis
physical	precipitate	reaction	respiration	substances	

1 The world around you is constantly changing. These changes can be classified as either ________________ or ________________ changes.

2 During a physical change, no new ________________ are produced.

During a chemical change, new substances are produced. A chemical change can be identified by a permanent change in ________________, a ________________ being given off, a ________________ forming from two clear solutions or energy being produced or absorbed in the form of ________________ and light.

3 Physical and chemical changes can be understood by the ________________ model.

This model assumes that all substances are made up of hard, indivisible particles. The model also assumes that these particles are ________________ to each other and are constantly ________________ .

4 When a chemical change occurs, scientists say that a chemical ________________ has taken place. This is the process of converting substances into different substances. During a chemical reaction the atoms rearrange to form new substances.

5 Chemical reactions are happening all around you all the time and are important to your daily life, industry and the environment. Combustion is one type of chemical reaction where substances burn with ________________ to create heat and light.
This chemical reaction is important for the burning of fossil ________________, which are used to produce ________________.

6 Chemical reactions are also important in biology. The chemical reaction known as ________________ allows plants to convert energy from the Sun into food energy. The chemical reaction known as ________________ is what allows plants and animals to convert food energy into the energy they need to live, grow, move and reproduce.

7 There is also chemistry occurring in the non-biological environment. Chemical weathering is one example, where rainwater (which is slightly ________________) dissolves rocks, statues and buildings.

RATE MY UNDERSTANDING
Shade the face that shows your rating

7.11 Thinking about my learning

1 Complete the table by stating three examples of a physical change and three examples of a chemical change.

Examples of a physical change	Examples of a chemical change
______________________	______________________
______________________	______________________
______________________	______________________

2 Use the terms you have learned related to changes of state to complete the sentences.

If ice is heated it ______________ and becomes water; more heat will result in the water boiling to become ______________. When the ______________ cools it ______________ to water which can then be ______________ to make ice.

Solid carbon dioxide, when heated does not become a liquid but turns directly into a ______________ in a process called ______________.

3 Circle the products and underline the reactants in the chemical reaction below.

$$4Fe(OH)_2 + O_2 + 2H_2O \longrightarrow 4Fe(OH)_3$$

4 Complete the table by placing a tick in the column you best think rates your understanding.

	Well	A bit	Not really
I feel I know the difference between a physical and a chemical change.			
I understand changes of state and know all of the vocabulary.			
I understand density.			
I understand chemical reactions can be represented by equations.			
I know the difference between a reactant and a product.			

CHAPTER 8

Rocks and mining

8.1 Knowledge preview

Science understanding

FOUNDATION	STANDARD	ADVANCED

1 Make a concept map using as many of the following terms as you can.

ash	crystal	deposition	diamond	erosion	foliation
fossil	granite	igneous	lava	limestone	magma
metamorphic	rocks	sediment	sedimentary	volcano	weathering

2 Select three of the words you have in the concept map and write a sentence for each showing that you understand what it means.

3 List four minerals that are mined in Australia.

4 Name two different methods of mining and briefly describe each method.

8.2 Weathering experiment

Science inquiry skills

FOUNDATION	**STANDARD**	ADVANCED

Communicating

Two students, Sally and Glenn, wanted to find out if physical and chemical weathering acted together in the wearing down of marble. They knew from previous experiments that acid could attack marble and limestone. They wanted to see if this chemical weathering was affected by physical weathering of the marble.

They designed an experiment where they smashed up some marble pieces into different sizes to represent physical weathering and added dilute hydrochloric acid to them. You can see the set up for their experiment below.

Sally and Glenn made sure they used the same mass of marble, 5 grams, and the same volume of hydrochloric acid, 25 mL, in each test-tube. In the first trial, Sally and Glenn used one piece of marble, in the second they used a number of large lumps, in the third trial they used numerous small lumps and in the final trial they used fine powder. As a control, they also set up 4 additional test-tubes with the same weight of marble but they used water instead of hydrochloric acid.

Physical and chemical wearing of marble

Liquid type	Size of marble (mass 5 g)			
	One piece	Large lumps	Small lumps	Fine powder
tubes with 25 mL acid				
tubes with 25 mL water				

marble (*n*) a smooth, hard rock formed from limestone

dilute (*adj*) a weak solution, a low amount of solute

They decided to measure the mass of marble left after 10 minutes to decide how fast the marble weathered. After 10 minutes, they tipped each of the test-tube contents through separate filter papers. They then rinsed the solid captured in each filter paper with water until they were sure there was no acid left. Then they dried and weighed each filter paper to see how much of the marble solid was left.

Their results are shown in the table below. This shows the mass of marble left after 10 minutes in acid or water. Remember that each test-tube originally had 5 grams of marble at the start of the experiment.

Physical and chemical wearing of marble results

Liquid in test-tube	Mass of marble left (g)			
	One piece	Large lumps	Small lumps	Fine powder
acid	4.5	1.1	0.5	0
water	5.0	5.0	5.0	5.0

8.2 Weathering experiment

1. State a hypothesis for Sally and Glenn's experiment.

2. Explain why Sally and Glenn used the test-tubes with water in them.

3. Explain why Sally and Glenn made sure they started with the same mass of marble in each test-tube.

4. Propose why they rinsed (washed) the filter paper with water.

5. Explain why they dried the filter paper before weighing it.

6. Interpret the results to produce a conclusion for this experiment.

7. Propose how Sally and Glenn could have improved on the design of this experiment.

RATE MY UNDERSTANDING
Shade the face that shows your rating  

8.3 Settling sediments

Science inquiry skills

FOUNDATION | STANDARD | ADVANCED

Processing & Analysing | Communicating

Sediments are carried by rivers and eventually deposited. When the river is flowing fast, large sediments can be carried. When the river slows, only smaller sediments can be carried. If the river stops flowing, then even the smallest sediments are deposited.

By studying the particles in sediment, you can learn a lot about the history of a place. The first step is to use a stack of sieves to pass the sediments through. The sieves are of different sizes, from large at the top to small at the bottom. These sieves sort out the grains into size ranges. The next step is to measure the mass of each category of grain size.

Look at the diagram to the right. Consider the sites A, B and C shown along the river valley.

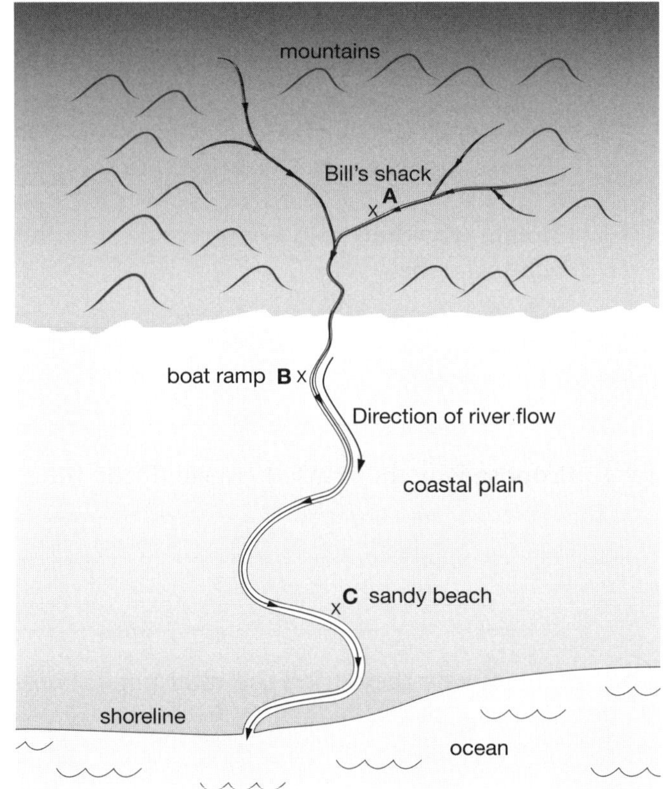

The river valley was studied by sampling the sediments at two of the sites, A (Bill's shack) and C (sandy beach). The results for sites A and C are shown in the table.

Sediment sample results

Particle size (mm)	Mass of particles (g) at site:	
	A: Bill's shack	C: sandy beach
8.00–16.00	320	0
4.00–8.00	560	0
2.00–4.00	1200	0
1.00–2.00	1410	0
0.50–1.00	2410	0
0.25–0.50	700	600
0.125–0.250	400	1050
0.050–0.125	0	2000
0.025–0.050	0	2600
Less than 0.025	0	750

8.3 Settling sediments

1. On the axes below, construct column graphs of these results. Note that two columns have already been drawn for site A to show you what to do. Write a heading for each graph.

Site A: Bill's shack

Mass of particles (g)

2500
2000
1500
1000
500

Less than 0.025 | 0.025–0.050 | 0.050–0.125 | 0.125–0.250 | 0.25–0.50 | 0.50–1.0 | 1.0–2.0 | 2.0–4.0 | 4.0–8.0 | 8.0–16.0

Particle size

Site C: Sandy beach

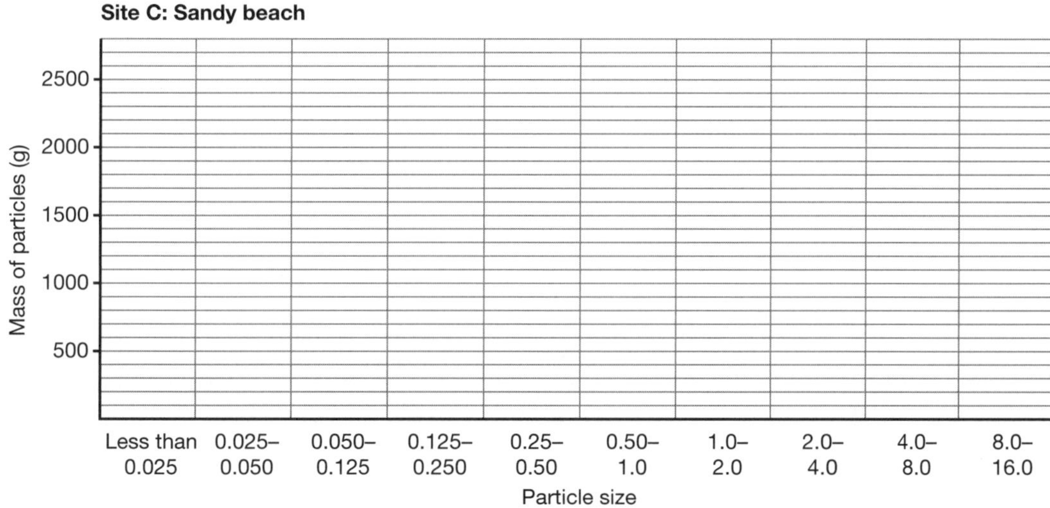

2. Identify the site where the sediment is mainly composed of smaller particles.

____

3. Discuss some differences in the range of the particle sizes at the two sites.

____

____

4. Propose why there was a difference in the average size of the particles found at sites A and C (about 1 mm compared with about 0.05 mm).

____

____

5. Predict what the particle sizes may be like at site B, the boat ramp. Justify your answer.

____

____

RATE MY UNDERSTANDING
Shade the face that shows your rating  

8.4 The Grand Canyon

Science understanding

FOUNDATION | STANDARD | **ADVANCED**

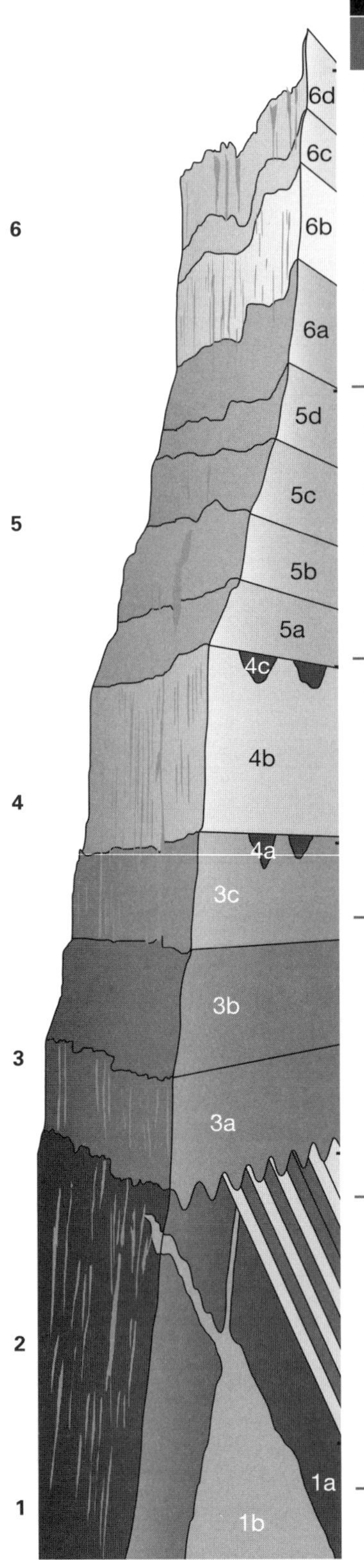

Figure 8.4.1 The Grand Canyon rock layers

The Grand Canyon rock layers and descriptions

Layers of rocks	Descriptions of layers
6 Hermit, Coconino, Toroweap and Kaibab	
6d Hermit shale	• Layer 6d has marine fossil-like shells so it must have been below the sea.
6c Coconino sandstone	• Layer 6c has some marine limestone, showing the sea covered the land again.
6b Toroweap formation	• Layer 6b is formed from sand dunes on land.
6a Kaibab limestone	• Layer 6a has fossils of winged insects, cone-bearing plants and ferns, amphibians and reptiles. The environment must have been swampy when rock formed.
5 Supai Group	
5d Watahomiai formation 5c Manakacha formation 5b Wescogame formation 5a Esplanande formation	• Contains fossils of land plants, amphibian (animals that live on land and in water such as frogs, newts) footprints and reptile (cold-blooded vertebrate animals such as snakes, turtles) fossils. • Shows the sea was retreating and land was forming as these rock layers were formed.
4 Temple Butte, Redwall and Surprise Canyon	
4c Temple Butte limestone	
4b Redwall limestone	• Layer 4b is thick limestone showing the area was under the sea. There are many fossils of sea creatures such as sea stars, sponges and corals.
4a Surprise Canyon formation	• Section 4a is limestone with fossils of marine and freshwater fish with backbones.
3 Tonto Group	• Three horizontal sedimentary layers.
3c Tapeats shale	• Layer 3c is a chemical sedimentary rock formed in a shallow sea.
3b Bright Angel shale	• Layer 3b is shale formed from mud that was just offshore and contains fossils of brachiopods, trilobites and worms.
3a Muav limestone	• Layer 3a contains fossils of sea creatures called trilobites (extinct relatives of crabs) and brachiopods (shelled sea creatures).
2 Grand Canyon Supergroup	• All layers have been tilted by the Earth's movements. • The top layer is mainly sedimentary layers formed in shallow seas. • Below that is a basalt layer from volcanic activity. • The sedimentary rock shale layer that has ripple marks that were formed in shallow sea. • Lowest sedimentary layer contains fossilised stromatolites formed by blue-green bacteria that lived in shallow coastal seas. The Grand Canyon was a shallow sea when this layer was formed.
1 Vishnu Group	• Mainly igneous rock
1a Vishnu schist	• Layers tilted at angle due to Earth movements
1b Zoroasterter granite	• No sedimentary rocks

8.4 The Grand Canyon

The Grand Canyon is one of the deepest cuts in the Earth's crust where rock layers are exposed and visible. It is found in the United States in Arizona, about 500 km from the sea. It was created by the Colorado River cutting its way down to a depth of 1.6 km into the rock. Geologists have concluded it took about 70 million years for the canyon to form. The canyon gave geologists an opportunity to study the sequence in which the rocks were laid down. They concluded that the oldest rocks at the bottom of the canyon are 2000 million years (2 billion years) old.

The names and types of rock found in the different strata are shown in Figure 8.4.1.

1 Identify the oldest sedimentary layer in the Grand Canyon.

____

2 Identify the youngest sedimentary layer in the Grand Canyon.

____

(3) Propose why geologists would think that the Redwall limestone was laid down over a very long period of time.

____

____

____

(4) Clastic sedimentary rocks are composed of weathered rock particles cemented together. Identify three layers that are clastic sedimentary rocks.

____

____

____

(5) The Grand Canyon is over 500 km inland from the sea. Discuss some evidence from these rock layers supporting the view that in the past this area was sometimes covered by the sea, but not at other times.

____

____

____

[6] Explain how the Grand Canyon is evidence that the Earth's sedimentary rocks were laid down over a very long period of time.

____

____

____

RATE MY UNDERSTANDING
Shade the face that shows your rating 

8.5 The rock cycle

Science understanding

FOUNDATION	STANDARD	ADVANCED

The rock cycle is the weathering and erosion of rocks and the way one type of rock forms into another type. Consider Figure 8.5.1—the rock cycle.

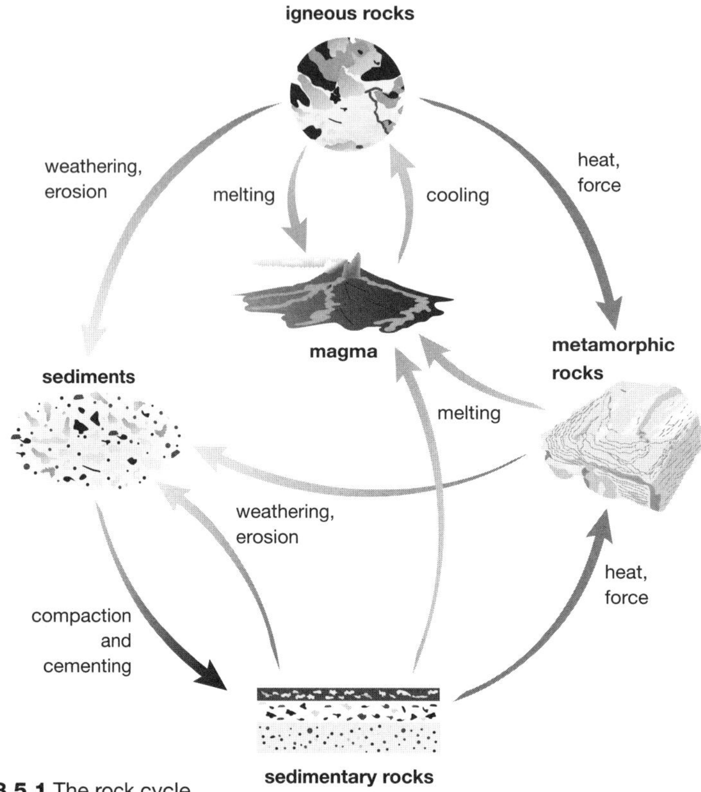

Figure 8.5.1 The rock cycle

1. Starting at magma in Figure 8.5.1, identify which type of rock forms first, and the process by which it forms.

____

____

2. Once an igneous rock forms, explain how it can be recycled to become an igneous rock again.

____

____

3. Identify the processes that turn an igneous rock into a sedimentary rock.

____

____

4. Identify the processes that turn an igneous rock into a metamorphic rock.

____

____

8.5 The rock cycle

5. Identify the process that changes metamorphic rocks into sedimentary rocks.

6. Identify the process that changes sedimentary rocks into metamorphic rocks.

7. Explain how a sedimentary rock can turn into an igneous rock.

8. Assess how useful the rock cycle is as a model in geology.

RATE MY UNDERSTANDING
Shade the face that shows your rating   

8.6 Igneous rocks

Science understanding

FOUNDATION | **STANDARD** | ADVANCED

Use your knowledge of igneous rocks and conduct further research to complete the following tasks.

1 **(a)** Figure 8.6.1 shows a variety of intrusive and extrusive volcanic features. The names of these features are listed below. Label the diagram using words from the box. You may use words more than once.

ash	batholith	cone	crater	dyke	laccolith
lava	magma	magma chamber	parasite cone	sill	vent

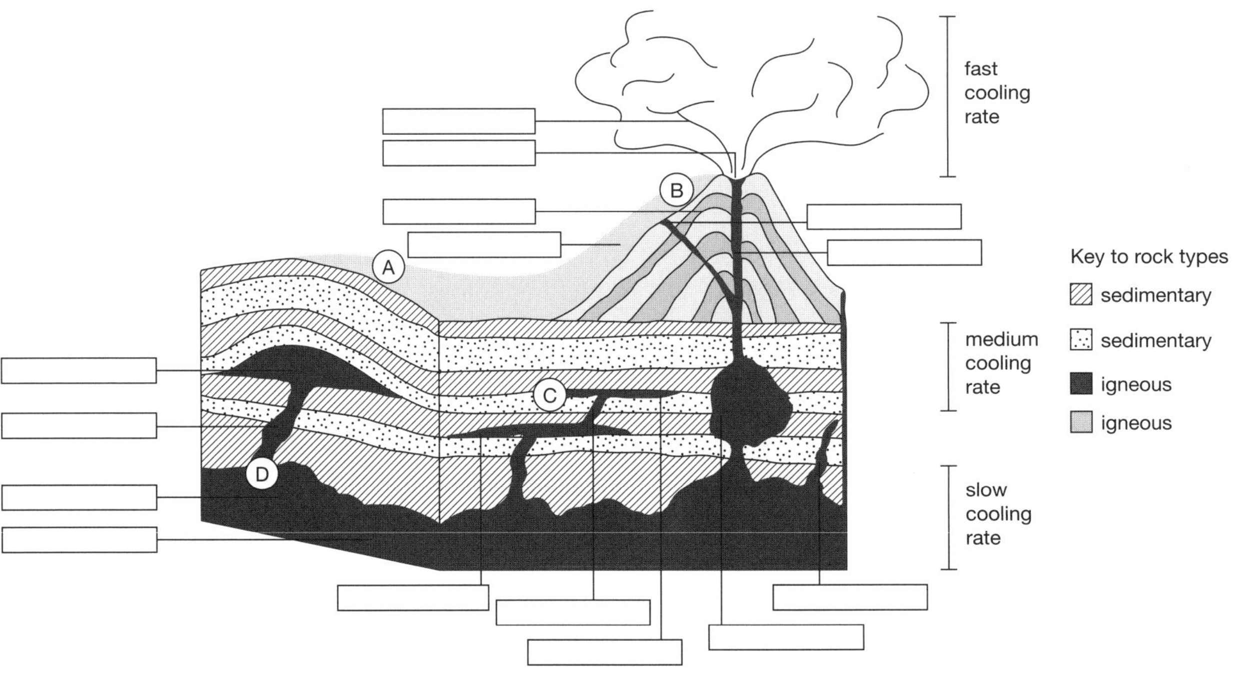

Figure 8.6.1 Intrusive and extrusive volcanic features

(b) Describe each of the following volcanic features:

(i) magma ______________________________

____

(ii) laccolith ______________________________

____

(iii) crater ______________________________

____

(iv) sill ______________________________

____

(v) dyke ______________________________

____

(vi) lava ______________________________

____

(c) If you looked at the soil near locations A and B, what difference do you think you would find in the grains (rock particles)?

____

____

(d) Explain why this difference would occur in the soil particles at locations A and B.

____

____

(e) Igneous rock samples were taken from locations B, C and D. Identify which rocks are intrusive igneous rocks and which are extrusive igneous rocks.

____

____

2 Read the descriptions of the igneous rocks provided below.

	Intrusive igneous	Extrusive igneous
Coarse grains or crystals	diorite, gabbro, granite	
Fine grains or crystals		basalt, rhyolite
Air holes in rock		pumice, scoria
No crystals		obsidian

(i) Name two course-grained igneous rocks.

____

(ii) Identify where course-grained igneous rocks form.

____

(iii) Explain why course-grained igneous rocks form at this location.

____

____

(iv) Rhyolite and granite have a similar mix of minerals, yet rhyolite has finer crystals forming than granite. Account for the difference in their crystal sizes.

(v) Explain why no crystals form in obsidian.

(vi) Explain how pumice, a volcanic rock, can float on water.

3 Lava or volcanic bombs can vary from 65 cm to 6 m in length They are extrusive yet have both very fine crystals and very coarse crystals as shown in Figure 8.6.2 where a lava bomb is split open. Explain how lava bombs form different-sized crystals.

Figure 8.6.2 A lava bomb

RATE MY UNDERSTANDING
Shade the face that shows your rating

8.7 Metamorphic changes in rocks

Science understanding

FOUNDATION	STANDARD	**ADVANCED**

Metamorphic rocks are formed by heat and pressure. The amount of heat and pressure experienced by a rock changes its final appearance and characteristics. Figure 8.7.1 lists how different metamorphic rocks are formed.

Figure 8.7.1 Metamorphic changes in rocks

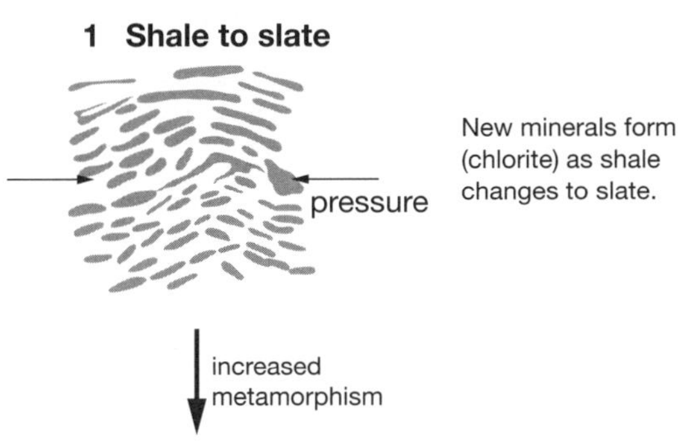

New minerals form (chlorite) as shale changes to slate.

With a fairly low amount of pressure, and at fairly low temperatures of a few hundred degrees, the sedimentary rock shale changes into slate, a type of metamorphic rock. The clay minerals in shale have changed into different minerals such as chlorite. The crystals in slate grow into microscopic flat shapes. Slate splits easily in the direction of the flattened crystals.

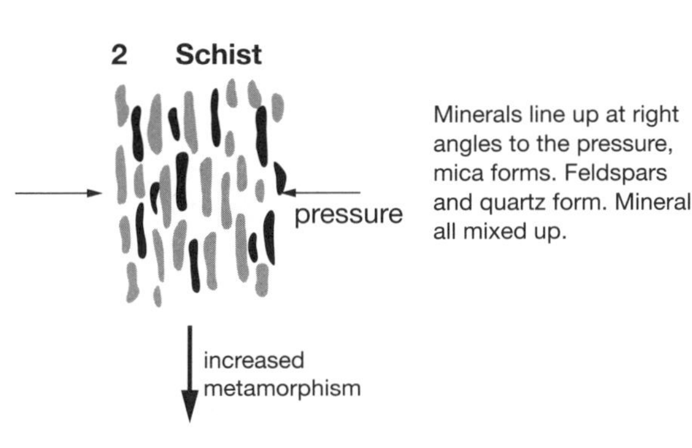

Minerals line up at right angles to the pressure, mica forms. Feldspars and quartz form. Minerals all mixed up.

If the slate is then subjected to even more pressure, and higher temperatures, it changes again into a schist. The crystals are moved around and line up at right angles to the direction of the pressure on the rock. New minerals form such as micas, quartz and feldspars. The different minerals are still all mixed together. The crystals grow large enough to be seen and become flattened and form sheets. The sheets become flaky and the rock is easily split between the sheets.

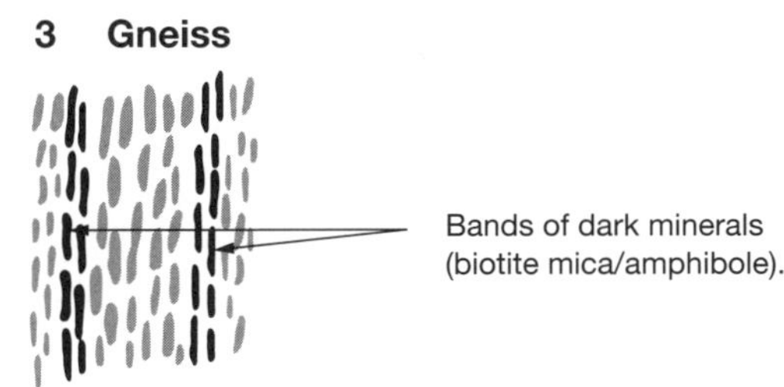

Bands of dark minerals (biotite mica/amphibole).

With even more pressure, and increased temperature, the schist changes again into a gneiss. The crystals may form into bands. For example, crystals of the dark coloured minerals biotite and amphibole squash together into a dark coloured band. All the quartz and feldspar crystals line up into a light coloured band. The direction of the line of crystals is at right angles to the pressure that compressed the rock.

8.7 Metamorphic changes in rocks

1 Where does the pressure come from that changes shale into slate?

2 What is the difference between shale and slate?

3 What does slate become if it is subjected to pressure?

4 What differences are there between slate and schist?

5 Why are there dark bands in a gneiss?

6 In what direction was the pressure acting to form the bands in a gneiss?

RATE MY UNDERSTANDING
Shade the face that shows your rating

8.8 Identifying rocks

Science inquiry skills

FOUNDATION	STANDARD	ADVANCED

Processing & Analysing | Communicating

1. Consider the diagrams of rocks shown in diagrams A to F. Classify each of the rocks and justify your decision. Select the type of rock from the box and write your answer in the table below. Two of the answers have been completed for you.

banded metamorphic	clastic sedimentary (breccia)	clastic sedimentary (sandstone)
extrusive igneous	intrusive igneous	organic sedimentary

A

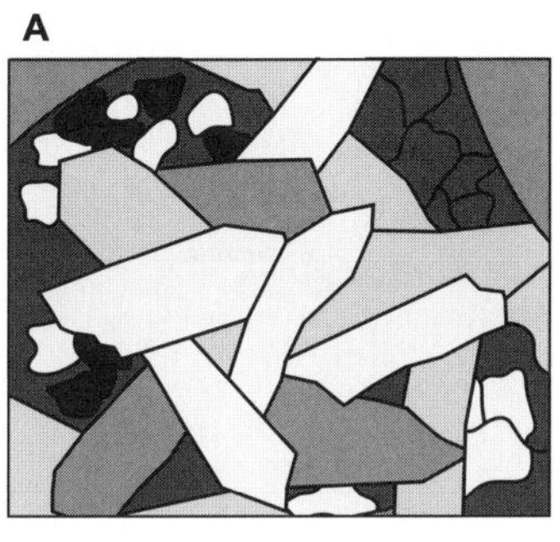

B

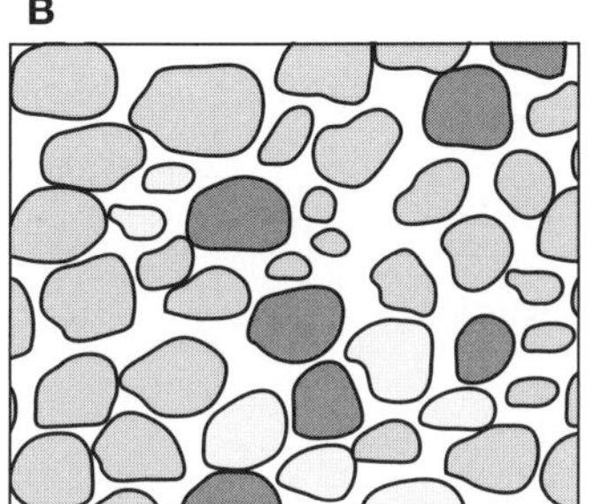

C

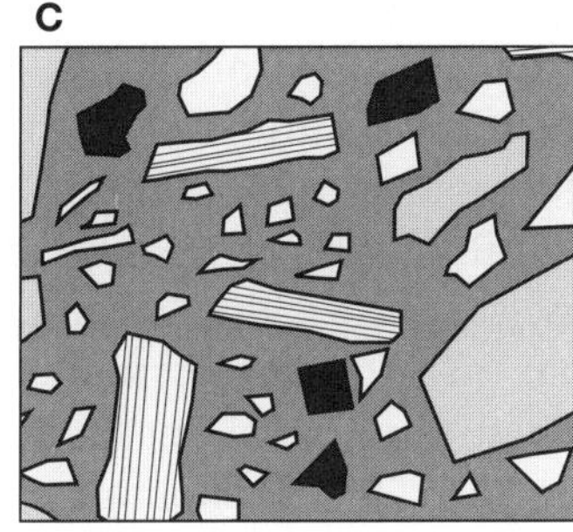

D

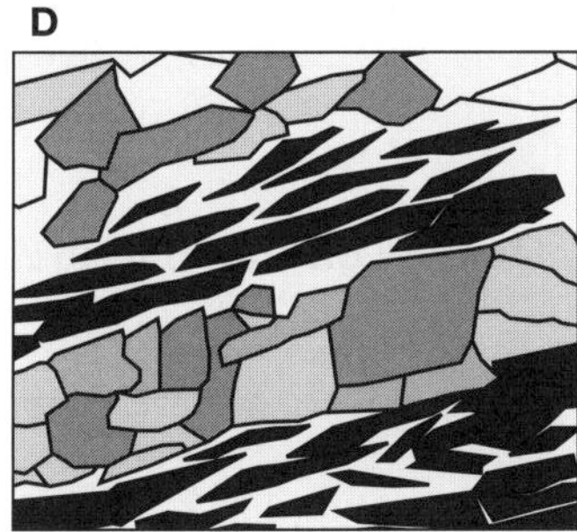

E

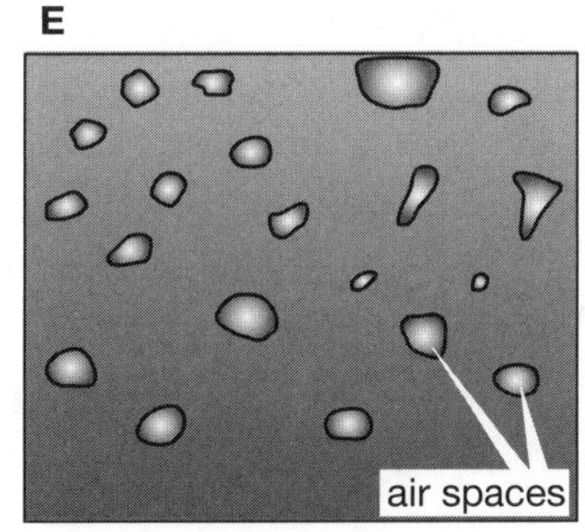

F

Rock	Type of rock	Justification for classification
A	large interlocking crystals	
B		
C		
D	banded metamorphic	
E		
F		

RATE MY UNDERSTANDING
Shade the face that shows your rating

8.9 The Super Pit

Science as a human endeavour

FOUNDATION	STANDARD	**ADVANCED**

The Firmiston Open Pit (Figure 8.9.1) is in the Golden Mile near Kalgoorlie-Boulder in Western Australia. This area is one of the richest gold deposit areas the world. Since the discovery of gold here in 1893, the Golden Mile has produced over 1 644 270 kilograms of gold. In the early 1900s there were about 50 small mining companies operating that had dug hundreds of shafts for underground mining of gold and 3000 kilometres of underground tunnels. In the 1980s large companies began buying out the smaller companies to carry out large scale operations. The Firmiston Open Pit was a result of the joining of many smaller pits. This pit is so large it is called the Super Pit.

Conduct research into the Super Pit using the Internet.

1 **(a)** Identify the location of the Super Pit by labelling Kalgoorlie on the map of Australia above. Label where you live on the map.

(b) Describe the location of the Super Pit in relation to Perth and to where you live. You might consider direction and distance.

____

____

2 State the length, width and depth of the Super Pit.

Length: ____________

Width: ____________

Depth: ____________

Figure 8.9.1 The Firmiston Open Pit—the Super Pit

(3) Look up images (photos and maps) of the Super Pit on the Internet. Your search terms could be 'Super Pit mining Western Australia'. Describe your observations by completing the Y-chart.

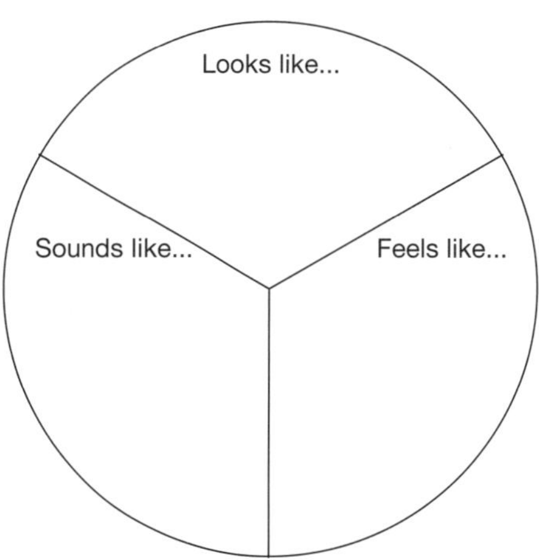

(4) Mining operations at the Super Pit follow a series of steps. Read the jumbled description of the steps. Identify the order of mining operations, numbering them 1 to 9 in the spaces provided.

Mining operations steps		
Step ______________ Mining proceeds on a series of levels, called benches. Geologists locate the ore blocks on the bench by drilling holes and checking the quality of the ore to find out how much gold is in each sample.	Step ______________ The Super Pit is built over many old underground mines. The old workings are marked with flags for safety.	Step ______________ Workers and equipment are not allowed into a blast site for at least 12 hours after the blast. This gives the ground time to settle.
Step _______ Ore dug out of the mine is transported to a processing plant to begin extraction of gold.	Step__________ Geologists mark where the ore blocks are with pegs and tape, to identify them for the digging teams. The digging crew follow the information from geologists and start removing ore.	Step__________ Ore is crushed and ground into smaller and smaller sized pieces. Extracted gold is melted and put in moulds to make gold bars. The gold is 60–80% pure.
Step _________ People in nearby towns are informed of blast times, and can see the process from a viewing platform.	Step ____________ Each gold bar is stamped with its own unique number. The gold bars are transported to the Australian Gold Refinery in Perth to be refined into 99.9% pure gold.	Step __________ The hard rock is blasted apart with explosives. Holes are drilled in a pattern about 5 metres apart and filled with the explosive ammonium nitrate and fuel oil or diesel at the bottom. The hole is filled with several metres of gravel to force the explosion into the rock, not back out the hole.

5 Explain why old underground workings are a problem at the Super Pit.

____

____

6 Explain why blasting holes are plugged at the top.

____

____

8.10 Literacy review

Science understanding

FOUNDATION	STANDARD	ADVANCED

1 Match the clues with the correct terms by ruling a line between them.

Clue	Term
The study of rocks, their history and the processes that form and change them	rock cycle
Crystals that grow into each other in a rock	mineral
Molten rock that does not reach Earth's surface	metamorphic
The physical and chemical processes that break rocks down into smaller pieces	fossils
Rocks made by sediments being cemented together	physical
The preserved evidence in rocks or soils of organisms that once existed on Earth	deposition
Layers of sedimentary rock	weathering
Process where minerals under pressure become compressed and the rock develops layers or bands	strata
Rocks formed when high temperature and pressure alter existing rocks	geology
The model geologists use to explain the endless cycle of change that happens to rocks as they change from one form to the other	igneous
The process of water or wind depositing eroded rock particles	extraction
Rocks formed from cooling magma	hydrothermal
A naturally occurring inorganic liquid or solid in Earth's crust	magma
Type of weathering caused by ice, salts, wind, water and temperature change	foliation
Chemical process that separates the wanted metal from the mineral in which the metal occurs	interlocking
Water heated above its boiling point and that moves through rocks in the crust	sedimentary

RATE MY UNDERSTANDING
Shade the face that shows your rating

8.11 Thinking about my learning

1 Construct a concept map using as many of the following terms as you can.

ash	crystal	deposition	diamond	erosion	foliation
fossil	granite	igneous	lava	limestone	magma
metamorphic	rocks	sediment	sedimentary	weathering	volcano

(2) Add at least 5 extra terms that you have learned in the chapter on rocks and mining to your concept map in question 1.

3 Compare the concept map on this page with the one you made at the beginning of the unit in Worksheet 8.1. Describe the similarities and differences between the two concept maps.